AF703282

AGRICULTURAL METEOROLOGY AND CLIMATOLOGY

AGRICULTURAL METEOROLOGY AND CLIMATOLOGY

PRINCIPLES AND APPLICATIONS

S N Pandey
Hari Maurya

Kruger Brentt
Publishers
2023

Kruger Brentt Publishers UK. LTD.
Company Number 9728962

Regd. Office: 68 St Margarets Road, Edgware, Middlesex HA8 9UU

© 2023 AUTHORS

ISBN: 9781787150379

Disclaimer:

Every possible effort has been made to ensure that the information contained in this book is accurate at the time of going to press, and the publisher and author cannot accept responsibility for any errors or omissions, however caused. No responsibility for loss or damage occasioned to any person acting, or refraining from action, as a result of the material in this publication can be accepted by the editor, the publisher or the author. The Publisher is not associated with any product or vendor mentioned in the book. The contents of this work are intended to further general scientific research, understanding and discussion only. Readers should consult with a specialist where appropriate.

Every effort has been made to trace the owners of copyright material used in this book, if any. The author and the publisher will be grateful for any omission brought to their notice for acknowledgement in the future editions of the book.

All Rights reserved under International Copyright Conventions. No part of this publication may be reproduced, stored in a retrieval system, or transmitted in any form or by any means, electronic, mechanical, photocopying, recording or otherwise without the prior written consent of the publisher and the copyright owner.

For information on all our publications visit our website at http://krugerbrentt.com/

PREFACE

Agrometeorology, abbreviated from agricultural meteorology, puts the science of meteorology to the service of agriculture, in its various forms and facets, to help with the sensible use of land, to accelerate the production of food, and to avoid the irreversible abuse of land resources. Agrometeorology is also defined as the science investigating the meteorological, climatological, and hydrological conditions that are significant to agriculture owing to their interaction with the objects and processes of agriculture production. The definition of biometeorology adopted by the International Society of Biometeorology (ISB) states, "Biometeorology is an interdisciplinary science dealing with the application of fields of meteorology and climatology to biological systems" (Hoppe, 2000, p. 383). The general scope includes all kinds of interactions between atmospheric processes and living organisms—plants, animals, and humans. By this definition, it becomes evident that there are roughly three subbranches of biometeorology: plant, animal, and human biometeorology. The domain of agrometeorology is the plant and animal subbranches. The third subbranch, human biometeorology, is outside the scope of agrometeorology.

The present book entitled "***Agricultural Meteorology and Climatology: Principles and Applications***" contains fifteen chapters covering all related disciplines. These chapters include Introduction to agricultural meteorology, atmospheric pressure and its variation, weather forecasting, meteorological instruments, weather hazards and its impact, environmental physics, measurement of soil moisture, agro-climatic zones and climatic normal, animal climatology, applications of meteorology to agriculture, weather and agriculture relations, climate change and agriculture, wind and agrometeorology, observation and measurement of cloud and modifications of crop microclimate. Related terminology is given at the end for ready reference. It is an essential resource for students, city planners, and all who are concerned with the nature of contemporary urban development problems.

We are grateful to all those persons as well as various books, manuals, periodicals, magazines, journals etc. that helped in the preparation of this book. In spite of the best efforts, it is possible that some errors may have occurred into the compilation and editing of the book. Further queries, constructive suggestions and criticisms for the improvement of the book are always welcome and shall be thankfully acknowledged.

S N Pandey
Hari Maurya

CONTENTS

1

INTRODUCTION TO AGRICULTURAL METEOROLOGY

1.1 METEOROLOGY

Meteorology is a combination of both physics and geography Meteorology is a combination of both physics and geography. This science utilizes the principles of Physics to study the behaviour of air. It is concerned with the analysis of individual weather elements for a shorter period over a smaller area. In other words, the physical state of the atmosphere at a given place and time is referred to as "weather". The study of weather is called 'meteorology'. It is often quoted as the "physics of atmosphere".

Agricultural meteorology has advanced during the last 100 years from a descriptive to a quantitative science using physical and biological principles. The agricultural community is becoming more aware that using climate and weather information will improve their profitability and this will no doubt increase the demand for agrometeorological services. Hence it is timely that the needs and perspectives for agrometeorology in the 21st Century are grouped under two major headings: agrometeorological services for agricultural production and agrometeorological support systems for such services. Emphasis must be placed on the components of such support systems comprising of data, research, policies and training/ education/extension.

1.2 AGRICULTURAL METEOROLOGY

- A branch of applied meteorology which investigates the physical conditions of the environment of growing plants or animal organisms
- An applied science which deals with the relationship between weather/climatic conditions and agricultural production.
- A science concerned with the application of meteorology to the measurement and analysis of the physical environment in agricultural systems. The word 'Agro meteorology' is the abbreviated form of agricultural meteorology.

- To study the interaction between meteorological and hydrological factors on the one hand and agriculture in the widest sense, including horticulture, animal husbandry and forestry on the other (WMO).

Meteorology vs. Agricultural Meteorology

Meteorology	Agricultural Meteorology
Branch of atmospheric physics	Branch of applied meteorology or a branch of agriculture as it deals with agriculture
It is a weather science	It is a product of agriculture and meteorlgy
It is a physical science	It is a biophysical science
It aims at weather forecasting	It aims at improving quantity and quality of crop production through meteorological skills
Weather service is the concern	Agro advisory service to the farmers is the concern based on weather forecast
It is a linking science to the society	It is a linking science to the farming community

1.3 SCOPE OF AGRICULTURAL METEOROLOGY

For optimum crop growth, specific climatic conditions are required. Agrometeorology thus becomes relevant to crop production because it is concerned with the interactions between meteorological and hydrological factors on the one hand and agriculture, in the widest sense including horticulture, animal husbandry, and forestry, on the other (Figure 1). Its objective is to discover and define such effects and to apply knowledge of the atmosphere to practical agricultural use. The field of interest of an agrometeorologist extends from the soil surface layer to the depth down to which tree roots penetrate. In the atmosphere he or she is interested in the air layer near the ground in which crops and higher organisms grow and animals live, to the highest levels in the atmosphere through which the transport of seeds, spores, pollen, and insects may take place. As new research uncovers the secrets of meteorological phenomena, there is increasing interest in remote sensing and interactions between oceans and the atmosphere in shaping seasonal conditions.

1.4 IMPORTANCE OF AGRICULTURAL METEOROLOGY

The dangers to the natural resource base, crops, and livestock that have a meteorological component include pollution of soil and air; soil erosion from wind or water; the incidence and effects of drought; crop growth; animal production; the incidence and extent of pests and diseases; the incidence, frequency, and extent of frost; the dangers of forest or bush fires; losses during storage and transport; and all farm operations. Agrometeorology offers practical solutions for harnessing climate potential and for protection against or avoidance of climate-related risks.

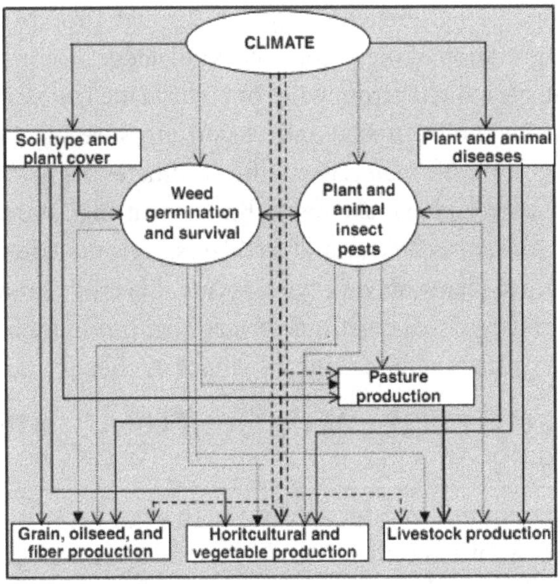

Fig. 1: Climate and agricultural production

The role of agrometeorol+ogy is both strategic and tactical. The strategic role is involved in the assessment of long-term utilization of natural resources in the development of crop diversity. The tactical role is more concerned with the short-term and field-scale decisions that directly influence crop growth and development. If communicated to the right client and applied, agrometeorological information can help farmers practice sustainable, high-quality, more profitable agriculture, with fewer risks, lower costs, and less environmental pollution and damage.

Of the total annual crop losses in world agriculture, a large percentage is due to direct weather effects such as drought, flash floods, untimely rains, frost, hail, and storms. Losses in harvest and storage, as well as those due to parasites, insects, and plant diseases, are highly influenced by the weather. When specifically tailored weather information is readily available to the needs of agriculture, it greatly contributes toward making short-term adjustments in daily agricultural operations, which minimize losses resulting from adverse weather conditions and improve the yield and quality of agricultural products.

Tailored weather information also provides guidelines for long-range or seasonal planning and the selection of crops most suited to anticipated climatic. Most decisions in livestock enterprises involve a considerable lag between decisions and their effects. Some decisions affect the product three to four years in the future. A long-range forecast is a very good climate risk-management tool which helps increase livestock production. Seasonal climate forecasts can play an important role in shaping the economic policies of governments. For example, with a forecast of a major drought, economic growth would be less than expected. By taking serious note of the forecast, monetary policy could be relaxed to maintain growth targets.

Other applications of agrometeorology are through improvement in techniques based on sound interpretation of meteorological knowledge. These include irrigation and water allocation strategies; shelter from wind or cold; shade from excessive heat; antifrost measures, including the choice of site; antierosion measures; soil cover and mulching; plant cover using glass or plastic materials; artificial climates of growth chambers or heated structures; animal housing and management; climate control in storage and transport; and efficient use of herbicides, insecticides, and fertilizers. Agrometeorological models can be used in efficient land-use planning; determining suitable crops for a region; risk analysis of climatic hazards and profit calculations in farming; production or harvest forecasts; and the adoption of farming methods and the choice of effective farm machinery.

1.5 CHRONOLOGY OF AGRICULTURAL METEOROLOGY DEVELOPMENTS

Attempts to relate agricultural production to weather go back at least 2,000 years and are still evolving. The twentieth century can be termed a very progressive and fertile one in respect to meteorology and agrometeorology. Qualitative studies in the nineteenth century were followed by statistical analyses, then by microclimatic measurements, and most recently by modeling.

1.5.1 Quantification of Crop-Weather Relationships

Visual observations of the microclimate and its impact on crop plants have been going on for several centuries. However, measurement of the characteristics of the microclimate in laboratory and experimental fields was strengthened during the first half of the twentieth century. The role of water in soil climate was recognized, and the link between the physical properties of soil, heat exchange, and water movement was investigated. It is well recognized that year-to-year variations in yields and regional commodity production are associated with variations in climate. Efforts were made to describe this relationship through statistical analysis of the correlation of yields with monthly rainfall. These analyses were a first attempt to use statistics to describe the nature of the relationship between variable yields, production, and climate. Later, the relationship between yields and rainfall was studied using multiple correlation methods. Since the 1920s (with some refinements in techniques), correlation and regression analyses have become the favorite tools for describing yield-weather relations.

1.5.2 Energy Balance Quantification

The first half of the twentieth century saw great contributions toward the quantification of water loss and use by plants. Research studies on the measurement and analysis of energy fluxes above crops and on crop evapotranspiration were stimulated. It was in this period that Bowen proposed a method (Bowen ratio) of partitioning the energy used in evaporation and heating the air. Penman published a rational method for using meteorological observations to estimate evaporation from a free water surface and vaporization from a plant canopy. The Penman method for estimating evapotranspiration has become a standard tool for estimating the need for irrigation water by agricultural engineers, agronomists, and meteorologists throughout the world. At the same time, the Thornthwaite method for estimating potential evapotranspiration was published. Further research into the energy balance resulted in the development of the eddy correlation method for estimating latent and sensible heat transfer. Later, Blaney and Criddle developed the consumptive use principle for irrigation scheduling. This technique has been widely used by agriculturists in the semiarid regions of the world. In more recent years, advanced and reliable instrumentation has made possible continuous measurements of biometeorological exchange processes, such as measurements of mass and energy exchange to assess the plant community's response to atmospheric variables.

1.5.3 Biological Studies in Controlled Climates

By the middle of the twentieth century, technology was available to build facilities in which biological responses to environmental conditions could be measured quantitatively. These facilities provided a way to study the responses of plants and animals to diurnal variations in weather conditions. One such facility was at the California Institute of Technology. At about the same time, a large animal facility at the University of Missouri, called the Missouri Climatic Laboratory, was established for studies dealing with the physiological response and production of dairy cattle to variations in temperature, humidity, wind, and radiation loads. As a result of these facilities, several excellent studies have contributed to a better understanding of climate and weather effects on plants and animals. These two laboratories served as precursors to the development of the growth chambers used today in nearly every part of the world.

1.5.4 Modeling Biological Response to Climate

By the late 1960s and early 1970s, an extensive literature was available that documented the response of plant growth and development to environmental conditions. This information paved the way for work on mathematical models of plant response and yields to varying environmental conditions. The comprehensive development and use of plant and animal dynamic simulation models started with the availability of computers in the early 1970s. By the close of the twentieth century, several thousand computer-based plant and animal dynamic simulation models had been developed to expand scientific insights into complex biological and environmental systems, and their use has resulted in huge economic benefits. The application of crop simulation models and simulation-based decision support systems became more acceptable to the agricultural community during the final decade of the twentieth century . Increases in the sophistication of computers and decreased costs are further fueling rapid advances in modeling. Interest has arisen in the topic of scaling from the leaf level to the global scale, to examine the global nature of climate change and its impact.

1.5.5 Remote Sensing of the Environment and Vegetation

Remote sensing detects and measures the characteristics of a target without being in physical touch with it. Information about the object is derived through electromagnetic energy. Aircraft and satellites are the main platforms for remote-sensing observations. Aerial photographs are the original form of remote sensing and remain the most widely used method. Infrared thermometry provides a way to determine the surface temperatures of plants and animals. Precise handheld infrared thermometers are commercially available to provide these measurements. The technology allows the measurement of the surface temperature with a resolution of a few square centimeters. The development and deployment of earth satellites in the 1970s brought a revolution in remote sensing. Remote sensing now provides a sequence of reliable and irreplaceable information for agriculture planning and management.

1.5.6 Weather and Climate Information for Agriculture

The agriculture industry is the most sensitive to variability in weather and climate. Throughout the world, efforts have been made to provide agriculture with a specifically focused weather service. Most countries of the world have developed programs to provide agroclimatological services. Unfortunately, in many countries there is a lack of coordination and cooperation to link agencies representing agriculture and meteorology in their efforts to advise farmers of weather-related risk management. This lack of cooperation has adversely affected improvements and further development in agro-advisory services. Furthermore, due to a lack of financial support, the network of meteorological stations does not adequately cover various agrometeorological zones to meet potential needs. Conflicts within and between countries often halt the collection and exchange of weather data. This has a detrimental impact on projects in which analysis of weather and climate data is attempted.

1.5.7 Crop-Related Climate Data Archives

Early studies dealing with relationships between yields and climate were accomplished using limited climatic data and primitive computational procedures. The advent of the computer era saw the development of new methods for storing historical data. Computer programs are now available that can electronically archive huge amounts of climatic data. These archives have further enhanced the evaluations of weather and climate risk for agriculture. The World Meteorological Organization (WMO) supports the sharing of computerized climate data in all countries of the world.

1.5.8 Climate Change and its Impact

Over the past 100 years, human activities have significantly altered the earth's atmosphere. Increases in the concentrations of greenhouse gases have led to warming of the earth's surface. An accumulating body of evidence suggested that by the last decade of the twentieth century global warming had already made significant negative impacts in a large number of regions. The menace of global climate change became a central issue of investigation in the 1990s and beyond. The investigations considered the effects of global warming on individual plants, plant stands, and entire vegetation units from regional to global scales. The investigations were not confined only to plants; the impact on hydro-resources, livestock, insect pests, and diseases has also been investigated.

1.6 FUTURE NEEDS

1.6.1 Agroclimatological Database

The availability of a proper meteorological and agrometeorological database is a major prerequisite for studying and managing the processes of agricultural and forest production. Historical data and observations during the current growing season will play a critical role in increased applications of crop models and model-generated output by farmers, consultants, and other policy- and decision makers. A major and inevitable priority is to build a database of meteorological, phenological, soil, and agronomic information. The acquisition of pertinent climate and agrometeorological data, their processing, quality control, and archiving, and timely access and database management are important components that will make the information valuable to agrometeorological research and operational programs. The major concerns for the availability of climate and agrometeorological data will continue to be in the areas of data collection and database management.

1.6.2 Agrometeorological Models

Agrometeorological models have many potentials uses for answering questions in research, crop management, and policy. As society becomes more computerized and technology oriented, there will be a greater possibility for the application of crop simulation models and decision support systems to help provide guidance in solving real-world problems related to agricultural sustainability, food security, the use of natural resources, and protection of the environment.

1.6.3 Environmental Management

A major area for future research is the response of environmentally sensitive agricultural practices to weather events. As the public becomes more concerned about the environment, greater pressure will be put on the agricultural community to document and prove that chemical applications are not harming the environment. This will require a better understanding of the role weather plays in the fate of agricultural chemicals during application, their persistence and movement after application, and their effect on natural organisms. To gain this understanding, research will require a more extensive interdisciplinary approach than is employed today. Adaptive research is required under on-farm or at least close to on-farm conditions, ideally with farmers participating.

1.6.4 Research

The most important development for science in general and for agrometeorology in particular is the rapid advances in electronics and their impact on computer, communication, and measurement technologies. The potential to handle data by computers and exchange them globally via the Internet is growing daily. Computers have opened the gates to the ability to store huge amounts of data and to process them through more computationally intensive statistical techniques (Serafin, Macdonald, and Gall, 2002). In agrometeorology, in which a vast amount of atmospheric data must be linked with complex sets of biological data, the availability of data in a uniform file format and the vanishing of data processing limitations result in a strong momentum for research.

1.6.5 Climate Change Impact

One of the most prominent current problems of humankind is global warming and its impact on the environment, water resources, agriculture, and human health. Agrometeorology has to play a leading role in the assessment of climate change, its impact on the biosphere, and adaptation strategies to increasing climate variability and climate change. Investigation of the effects of global warming on animals will be another challenge. Future animal agrometeorologists have to search deeper into animal responses to specifically defined factors of the environment. These findings will permit the development of more adaptive, more tolerant, and more productive animals in stressful environments.

1.6.6 Pest and Disease Management

Increasing environmental, population, and economic pressures are creating difficulties in solving agricultural pest and disease management problems. Future climate change and increased variability will further complicate pest and disease management problems. This will require improved analyses of the weather to develop new pest management techniques and strategies. Agrometeorologists trained in weather-pest and weather-disease relationships and in the basics of pest management disciplines need to play a key role in developing pest and disease management strategies.

1.6.7 Education and Training

Neither education nor training is a one-time effort. The acquisition of knowledge and skills should be viewed as a continuous process throughout one's career. The need for continued training in agrometeorology was demonstrated by a survey on education and training requirements by the WMO. The study revealed that the national meteorological and hydrological services in many countries do not have adequately trained personnel. Further-more, in many countries there are neither facilities nor sufficient national resources available to train personnel in the home country or abroad. In the academic setting, there is a need for creative educational programs in schools. These programs will educate younger generations on the importance of agriculture and how weather affects the food supply. Development of such programs requires professionals with an understanding of the interactions of weather and climate with agriculture. Unfortunately, fewer professionals are being trained in this discipline because of limited independent programs or departments in universities. Future training of agricultural meteorologists at the university level will require cooperative efforts between the WMO and member countries. A second area of education that has been lacking is that of the public and agricultural producers. As points out that agricultural producers need to learn how to better use weather-driven models in their daily decision making. More important, they need to be taught how weather affects the various decisions they make and how their productivity or profit can be improved by using this information. The perception is that both long- and short-range forecasts are not sufficiently reliable to use in decision making (Jagtap and Chan, 2000). Research programs are needed to improve and quantify the reliability of forecasts and show how these forecasts can be used to improve decision making. Subsequently, extension programs are needed to transfer these findings to agricultural producers.

1.7 SCOPE OF AGRICULTURAL METEOROLOGY IN AGRICULTURE

The science of meteorology has a wide scope in agriculture to improve use of land to increase agricultural production. The weather and climate is the natural resource and considered as basic input for agriculture. Every plant process related with growth, development and yield of crop is affected by weather. Similarly, every farm operation such as ploughing, harrowing, land preparation, sowing, weeding, irrigation, manuring, spraying, dusting, harvesting, threshing, storage and transport of farm produce are affected by weather.

The field of interest of agro meteorology extends from the soil surface layer to depth up to which tree roots penetrate. In the atmosphere, it is interested in the air layer near the ground in which crops and higher organisms grow and animals live, to the highest levels in the atmosphere through which the transport of seeds, spores, pollen and insects may takes place. Most of the meteorological phenomena like rain, cloud, winds, thunder, lightening etc. take place within the height of 30- 35 km, therefore, the meteorologist may consider this height with special interest.

The scope of agricultural meteorology is to make use of the knowledge of meteorology for developing a stable and sustained agricultural production system and can be illustrated through the following few applications.

1.7.1 Characterization of Agricultural Climate for Determining Crop Growing Season

Solar radiation, air temperature, precipitation, wind, humidity etc. are important climatic factors on which the growth, development and yield of a crop depends. The study and effect of these elements can be used for planning, stability and management of the crops on weather basis. The agricultural operations can be arranged by studying the concepts of effect of weather elements. According to climatic condition, agricultural year can be divided into three seasons i.e. kharif , rabi and summer.

1.7. 2 Crop Planning for Stability in Production

To reduce risk of crop failure on climatic part (due to climatic factor), so as to get stabilized yields even under weather adversity, suitable crops / cropping patterns / contingent crop planning can be selected by considering water requirement of crop, effective rainfall and available soil moisture. Stability can be further increased by selecting short duration crops with low water requirement, drought resistance or tolerance etc. Contingent crop planning, in case of late start or early start of monsoon as well as for long dry spells can also be done.

1.7.3 Crop Management on Weather Basis

Management of crop involves various farm operations such as sowing, interculturing operations, fertilizer application, plant protection, irrigation scheduling, harvesting etc. can be carried out on the basis of specially tailored weather report. For this the use of operational forecasts available from agromet advisories is made e.g.

- Sowing of crops can be carried out on the basis of past normal climatic data.
- Interculturing operations are undertaken during dry spell forecasted.
- Fertilizer application is advised when rainfall is less than 1mm / day and wind speed is < 30km/hr and soil moisture is between 30 to 80%. d) Spraying / dusting is undertaken when there is no rainfall and wind speed is < 25km /hr.

1.7.4 Crop Monitoring

To check crop health and growth performance, suitable meteorological tools such as crop growth models, water balance technique or remote sensing etc. can be used.

1.7.5 Crop Modeling and Yield Forecasting

A model is a mathematical equation giving growth and yield as a function of climatic parameters or other relevant parameters. Suitable crop models devised for the purpose can provide information or predict the results about the growth and yield when the current and past weather data are used.

1.7.6 Research in Crop-Climate Relationship

Agrometeorology can help to understand crop-climate relationship so as to resolve complexities of plant process in relation to its microclimate. It also develops suitable techniques of modification of microclimate.

1.7.7 Climatic Extremities and Management Strategies

Climatic extremities such as frost, floods, droughts, high winds, and hail storms can be forecasted. The loss in production can then be prevented / protected by arranging suitable strategies such as:

- ***Avoidance:*** e.g. Give up or No spraying if heavy rains are forecasted.
- ***Protection:*** e.g. Give irrigation to crop when frost is forecasted.
- ***Mitigation:*** e.g. small fire in garden when fog is forecasted. Mulching, interculturing operations if drought is forecasted.

1.7.8 Climate as a Tool to Diagnose Soil Moisture Stress

Soil moisture can be exactly determined from climatic water balance method which is used to diagnose the soil moisture stress, drought and necessary protective measures such as irrigation, mulching, application of antitranspirants, defoliation, thinning etc. can be undertaken.

1.7.9 Livestock Production

Livestock production is a part of agriculture. The set of favorable and unfavorable weather conditions for growth, development and production of livestock is studied in Agricultural meteorology. Thus, to optimize milk production and poultry production the climatic normals are worked out and on that basis the suitable breeds can be evolved or otherwise can provide the congenial conditions for the existing breeds.

1.7.10 Soil Formation

Soil formation process depends on climatic factors like temperature, precipitation, humidity and wind etc. Thus, climate is a major factor in soil formation and development.

1.7.11 Forest Production and Fire Prevention

Meteorological information is needed to conserve forest wealth. Natural forest fire destroying this valuable wealth can be forecasted well in advance from weather information.

2

ATMOSPHERIC PRESSURE AND ITS VARIATION

2.1 ATMOSPHERIC PRESSURE

An atmosphere (atm) is a unit of measurement equal to the average air pressure at sea level at a temperature of 15 degrees Celsius (59 degrees Fahrenheit). One atmosphere is 1,013 millibars, or 760 millimeters (29.92 inches) of mercury. Atmospheric pressure drops as altitude increases. Essentially, atmospheric pressure is the force exerted at any given point on the Earth's surface by the weight of the air above that point. In short: the air that surrounds the Earth creates atmospheric pressure and this pressure is determined by the collective weight of air molecules. Air molecules at higher altitudes have fewer molecules pressing down on them from above and therefore experience lower pressure, while lower molecules have more force or pressure exerted on them by molecules piled on top of them and are more tightly packed together. When you go up into the mountains or fly high in an airplane, the air is thinner and the pressure is lower. The air pressure at sea level at a temperature of 59°F (15°C) is equal to one atmosphere (Atm), and this is the baseline reading for determining relative pressure. Atmospheric pressure is also known as barometric pressure because it is measured using a barometer. A rising barometer indicates increasing atmospheric pressure and a falling barometer indicates decreasing atmospheric pressure.

2.2 CAUSES OF CHANGES IN ATMOSPHERIC PRESSURE

Changes in air pressure are caused by differences in air temperature above the earth, and the temperature of an air mass is determined by its location. For example, air masses above oceans are typically cooler than air masses above continents. Air temperature differences create wind and cause pressure systems to develop. The wind moves pressure systems and these systems tend to change as they pass over mountains, oceans, and other areas.,17th-century French scientist and philosopher Blaise Pascal (1623–1662) discovered that air pressure decreases with height and that pressure-changes at ground level can be attributed to the daily weather. These discoveries are used to predict the weather today. Often, weather forecasters refer to high- or low-pressure areas moving toward particular regions in order to describe predicted conditions for those areas. As air rises in low-pressure systems, it cools and often condenses into clouds and precipitation, resulting in storms. In high-pressure systems, the air sinks toward the Earth and warms upward, leading to dry and fair weather.

2.3 PRESSURE CHANGES: AFFECT THE WEATHER

In general, a mercury barometer can let you know if your immediate future will see clearing or stormy skies, or little change at all, based only on atmospheric pressure.

Here are a few examples of how to interpret barometric readings:

- When the air is dry, cool, and pleasant, the barometer reading rises.
- In general, a rising barometer means improving weather.
- In general, a falling barometer means worsening weather.
- When atmospheric pressure drops suddenly, this usually indicates that a storm is on its way.
- When atmospheric pressure remains steady, there will likely be no immediate change in the weather.

2.4 MEASURING ATMOSPHERIC PRESSURE

The atmospheric pressure is the weight exerted by the overhead atmosphere on a unit area of surface. It can be measured with a mercury barometer, consisting of a long glass tube full of mercury inverted over a pool of mercury:

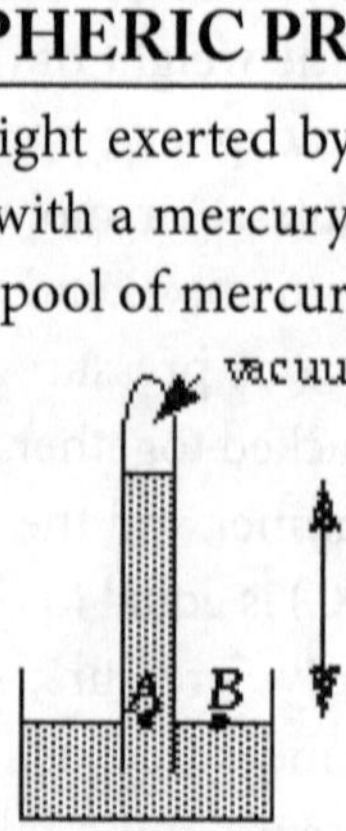

Fig. 1: Mercury barometer

When the tube is inverted over the pool, mercury flows out of the tube, creating a vacuum in the head space, and stabilizes at an equilibrium height h over the surface of the pool. This equilibrium requires that the pressure exerted on the mercury at two points on the horizontal surface of the pool, A (inside the tube) and B (outside the tube), be equal. The pressure PA at point A is that of the mercury column overhead, while the pressure PB at point B is that of the atmosphere overhead. We obtain PA from measurement of h:

$$P_A = \rho_{Hg} g h \qquad (1)$$

where rHg = 13.6 g cm^{-3} is the density of mercury and g = 9.8 m s^{-2} is the acceleration of gravity. The mean value of h measured at sea level is 76.0 cm, and the corresponding atmospheric pressure is 1.013x105 kg m^{-1} s^{-2} in SI units. The SI pressure unit is called the Pascal (Pa); 1 Pa = 1 kg m-1 s^{-2}. Customary pressure units are the atmosphere (atm) (1 atm = 1.013x105 Pa), the bar (b) (1 b = 1x105 Pa), the millibar (mb) (1 mb = 100 Pa), and the torr (1 torr = 1 mm Hg = 134 Pa). The use of millibars is slowly giving way to the equivalent SI unit of hectoPascals (hPa). The mean atmospheric pressure at sea level is given equivalently as P = 1.013x105 Pa = 1013 hPa = 1013 mb = 1 atm = 760 torr.

2.5 MASS OF THE ATMOSPHERE

The global mean pressure at the surface of the Earth is PS = 984 hPa, slightly less than the mean sea-level pressure because of the elevation of land. We deduce the total mass of the atmosphere ma:

$$m_a = \frac{4\pi R^2 P_S}{g} = 5.2x10^{18} kg \qquad (2)$$

where R = 6400 km is the radius of the Earth. The total number of moles of air in the atmosphere is Na = ma/Ma = 1.8x1020 moles.

2.6 VERTICAL PROFILES OF PRESSURE AND TEMPERATURE

Figure 2 shows typical vertical profiles of pressure and temperature observed in the atmosphere. Pressure decreases exponentially with altitude. The fraction of total atmospheric weight located above altitude z is P(z)/P(0). At 80 km altitude the atmospheric pressure is down to 0.01 hPa, meaning that 99.999% of the atmosphere is below that altitude. You see that the atmosphere is of relatively thin vertical extent. Astronomer Fred Hoyle once said, "Outer space is not far at all; it's only one hour away by car if your car could go straight up!"

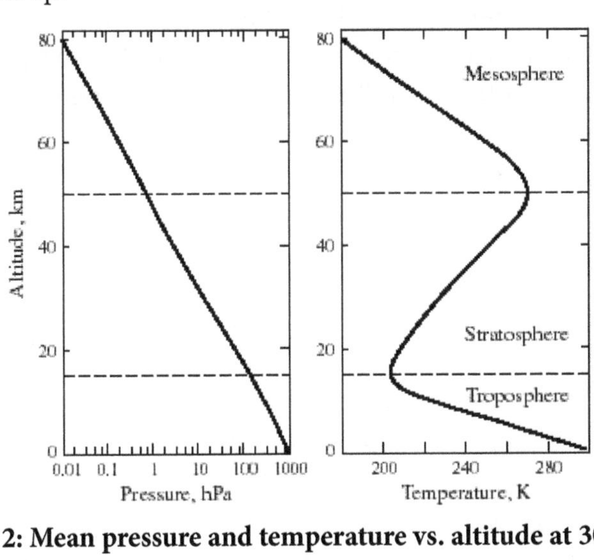

Fig. 2: Mean pressure and temperature vs. altitude at 30oN, March

Atmospheric scientists partition the atmosphere vertically into domains separated by reversals of the temperature gradient. The troposphere extends from the surface to 8-18 km altitude depending on latitude and season. It is characterized by a decrease of temperature with altitude which can be explained simply though not quite correctly by solar heating of the surface. The stratosphere extends from the top of the troposphere (the tropopause) to about 50 km altitude (the stratopause) and is characterized by an increase of temperature with altitude due to absorption of solar radiation by the ozone layer). In the mesosphere, above the ozone layer, the temperature decreases again with altitude. The mesosphere extends up to 80 km (mesopause) above which lies the thermosphere where temperatures increase again with altitude due to absorption of strong UV solar radiation by N2 and O2. The troposphere and stratosphere account together for 99.9% of total atmospheric mass and are the domains of main interest from an environmental perspective.

2.7 BAROMETRIC LAW

We focus here on explaining the vertical profile of pressure. Consider an elementary slab of atmosphere (thickness dz, horizontal area A) at altitude z:

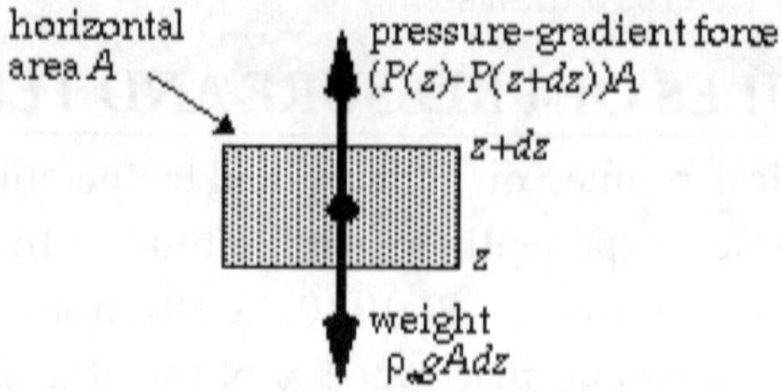

Fig. 3: Vertical forces acting on an elementary slab of atmosphere

The atmosphere exerts an upward pressure force P(z)A on the bottom of the slab and a downward pressure force P(z+dz)A on the top of the slab; the net force, (P(z)-P(z+dz)) A, is called the pressure-gradient force. Since P(z) > P(z+dz), the pressure-gradient force is directed upwards. For the slab to be in equilibrium, its weight must balance the pressure-gradient force:

$$\rho_a g A dz = (P(z) - P(z + dz))A \qquad (3)$$

Rearranging yields

$$\frac{P(z + dz) - P(z)}{dz} = -\rho_a g \qquad (4)$$

The left hand side is dP/dz by definition. Therefore

$$\frac{dP}{dz} = -\rho_a g \qquad (5)$$

Now, from the ideal gas law,

$$\rho_a = \frac{PM_a}{RT} \tag{6}$$

where Ma is the molecular weight of air and T is the temperature. Substituting (6) into (5)yields:

$$\frac{dP}{P} = -\frac{M_a g}{RT}dz \tag{7}$$

We now make the simplifying assumption that T is constant with altitude; as shown in Figure , T varies by only 20% below 80 km. We then integrate (7) to obtain

$$\ln P(z) - \ln P(0) = -\frac{M_a g}{RT}z \tag{8}$$

which is equivalent to

$$P(z) = P(0)\exp\left(-\frac{M_a g}{RT}z\right) \tag{9}$$

Equation (2.9) is called the barometric law. It is convenient to define a scale height H for the atmosphere:

$$H = \frac{RT}{M_a g} \tag{10}$$

leading to a compact form of the Barometric Law:

$$P(z) = P(0)e^{-\frac{z}{H}} \tag{11}$$

For a mean atmospheric temperature T = 250 K the scale height is H = 7.4 km. The barometric law explains the observed exponential dependence of P on z in Figure ; from equation (11) , a plot of z vs. ln P yields a straight line with slope -H (check out that the slope in Figure is indeed close to -7.4 km). The small fluctuations in slope in Figure are caused by variations of temperature with altitude which we neglected in our derivation.

The vertical dependence of the air density can be similarly formulated. From (2.6) , ra and P are linearly related if T is assumed constant, so that

$$\rho_a(z) = \rho_a(0)e^{-\frac{z}{H}} \tag{12}$$

A similar equation applies to the air number density na. For every H rise in altitude, the pressure and density of air drop by a factor e = 2.7; thus H provides a convenient measure of the thickness of the atmosphere.

In calculating the scale height from (10) we assumed that air behaves as a homogeneous gas of molecular weight Ma = 29 g mol-1. Dalton's law stipulates that each component of the air mixture must behave as if it were alone in the atmosphere. One might then expect different components to have different scale heights determined by their molecular weight. In particular, considering the difference in molecular weight between N2 and O2, one might expect the O2 mixing ratio to decrease with altitude. However, gravitational separation of the air mixture takes place by molecular diffusion, which is considerably slower than turbulent vertical mixing of air for altitudes below 100 km). Turbulent mixing thus maintains a homogeneous lower atmosphere. Only above 100 km does significant gravitational separation of gases begin to take place, with lighter gases being enriched at higher altitudes. During the debate over the harmful effects of chlorofluorocarbons (CFCs) on stratospheric ozone, some not-so-reputable scientists claimed that CFCs could not possibly reach the stratosphere because of their high molecular weights and hence low scale heights. In reality, turbulent mixing of air ensures that CFC mixing ratios in air entering the stratosphere are essentially the same as those in surface air.

The sea-breeze circulation

An illustration of the Barometric Law is the sea-breeze circulation commonly observed at the beach on summer days (Figure). Consider a coastline with initially the same atmospheric temperatures and pressures over land (L) and over sea (S). Assume that there is initially no wind. In summer during the day the land surface is heated to a higher temperature than the sea. This difference is due in part to the larger heat capacity of the sea, and in part to the consumption of heat by evaporation of water.

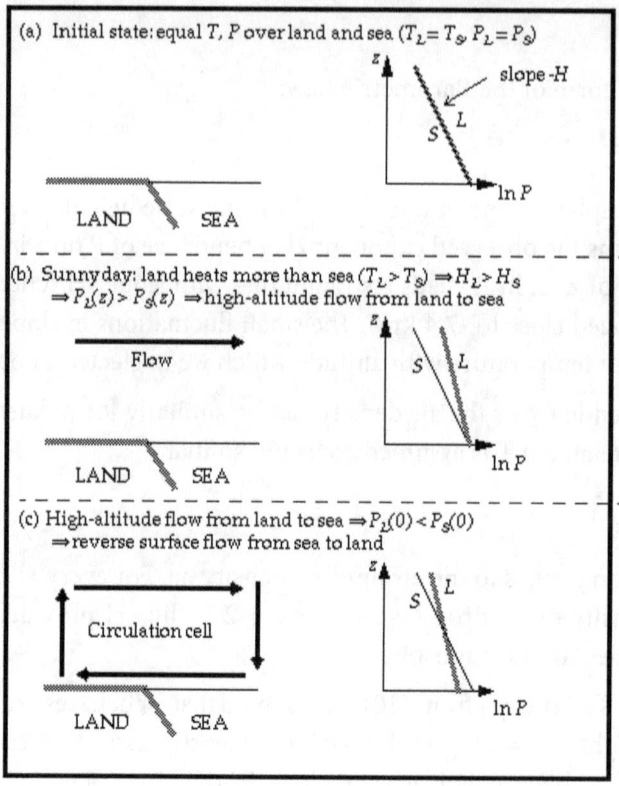

Fig. 4: The sea-breeze circulation

As long as there is no flow of air between land and sea, the total air columns over each region remain the same so that at the surface PL(0) = PS(0). However, the higher temperature over land results in a larger atmospheric scale height over land (HL > HS), so that above the surface PL(z) > PS(z) (Figure). This pressure difference causes the air to flow from land to sea, decreasing the mass of the air column over the land; consequently, at the surface, PL(0) < PS(0) and the wind blows from sea to land (the familiar "sea breeze"). Compensating vertical motions result in the circulation cell shown in Figure . This cell typically extends ~10 km horizontally across the coastline and ~1 km vertically. At night a reverse circulation is frequently observed (the land breeze) as the land cools faster than the sea.

Atmospheric pressure variation with height

- Pressure decreases with increasing altitude. The pressure at any level in the atmosphere may be interpreted as the total weight of the air above a unit area at any elevation. At higher elevations, there are fewer air molecules above a given surface than a similar surface at lower levels.
- For example, there are fewer molecules above the 50 km surface than are found above the 12 km surface, which is why the pressure is less at 50 km.

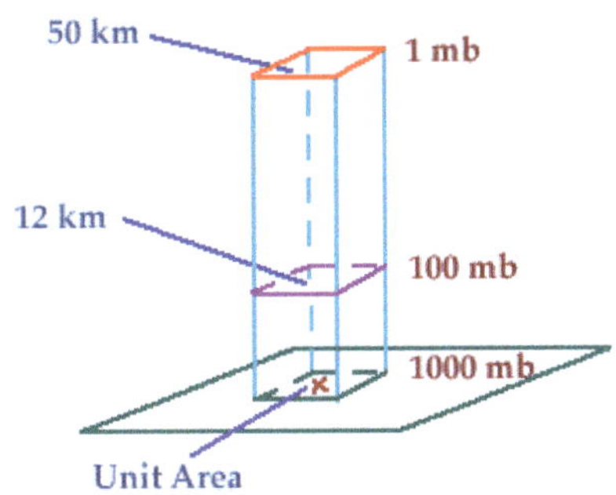

- What this implies is that atmospheric pressure decreases with increasing height. Since most of the atmosphere's molecules are held close to the earth's surface by the force of gravity, air pressure decreases rapidly at first, then more slowly at higher levels.

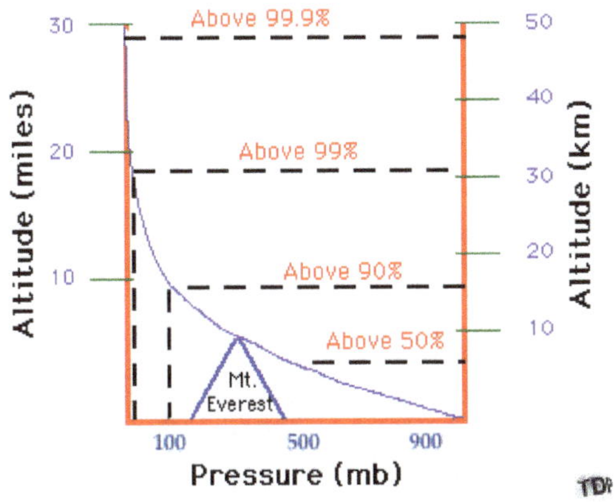

- Since more than half of the atmosphere's molecules are located below an altitude of 5.5 km, atmospheric pressure decreases roughly 50% (to around 500 mb) within the lowest 5.5 km. Above 5.5 km, the pressure continues to decrease but at an increasingly slower rate.

3

WEATHER FORECASTING

3.1. WEATHER FORECASTING

Weather affects nearly every one every day. Weather forecasts are issued to save lives, reduce property damage, reduce crop damage, to let the general public know what to expect. It is essential for especially for fishermen and farmers. Weather forecasting of what the weather will be like in an hour, tomorrow, next week or during a season. Surface observations of temperature, humidity, pressure, wind, visibility, clouds etc. are obtained by installing properly exposed instruments. Observations at various levels in the atmosphere are made by optical or radio/radar tracking of hydrogen-filled balloons carrying appropriate sensors. Radio sounding (radiosonde) and radio wind (RAWIN) observations provide measurements of pressure, temperature, humidity and winds at various levels such as also of atmospheric constituents such as ozone and carbon dioxide.

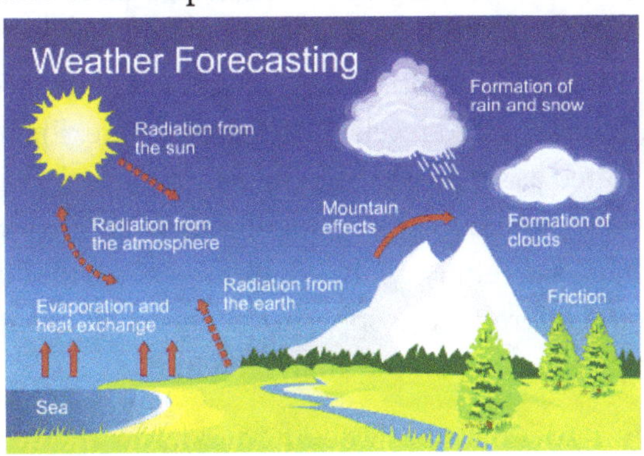

3.2. SYNOPTIC OBSERVATIONS

Simultaneous observations of the atmosphere at the surface and in the upper air are made at a large number of locations several times a day at specified hours all over the world. These observations are known as synoptic observations, are made by observers who scan the skies, read the instruments and faithfully record the observations. Surface observations are made every three hours and upper air observations, every six hours. Over and above these, various types of routine and special weather observations are made at different times to meet the specific requirements of various interests such as aviation, shipping, agriculture, hydrology etc. The data are fed into the sophisticated computers to interpret the weather conditions.

3.3. ORGANIZATION

The meteorological observations of all countries must be strictly comparable, so that they can be used and analyzed together. For this, standardization of instruments, observational procedures and data processing have to be ensured by international cooperation and agreement. This is made possible by the World Meteorological Organization (WMO) which is a specialized agency of the United Nations Organization. In India the Meteorological Department (IMD) provides climatological, weather forecasting and warning services for all interests, as also training and research facilities in respect of atmospheric sciences. IMD has forecasting offices at its 6 Regional Meteorological Centers and Metrological Centers at the State capitals. In addition, it has two major forecasting centers at New Delhi and Pune. IMD maintains a countywide network of observational stations from where weather reports are received at Pune through the Regional Telecommunication Hub based at New Delhi. With the help of these reports and other observations from the globe, the Pune office issues the All-India Daily Weather Bulletin and Weather Forecasts for the entire country.

3.4. SYNOPTIC CHARTS

A large volume of data is received at a weather office. These data have to be presented in an easily assimilable form. For this, the observations are plotted on a map and this map shows at a glance the weather conditions over a large geographical area at a specified time. Such a map is known as a synoptic chart. Charts are plotted for each synoptic hour for the surface and various standard levels in the atmosphere. Elements such as pressure, temperature, etc. are plotted as symbols at fixed positions around a circle representing the observing station. Wind is plotted as an arrow coming to the station from the direction it blows. The synoptic chart is carefully analyzed and the pattern of distribution of various parameters brought out. Different atmospheric systems such as lows, highs, depressions and cyclones are clearly delineated on the analyzed chart.

3.5. FORECASTS

Weather forecasts provide valuable guidance for forward planning, economic benefit as well as safety of men and materials. A forecast is a statement of weather conditions expected over a place or an area during a specified period of time in the future. Short range forecasts are valid up to forty-eight hours and give details of various weather elements and their variations expected during that period. Examples are the usual forecasts issued to the media and route forecasts issued to aircraft. Medium range forecasts are valid from two to seven days and are often tagged on to short range forecasts as further look out. Long range forecasts pertain to weeks, months or a season ahead and indicate the expected mean conditions of certain selected elements such as rainfall or temperature. The seasonal monsoon forecast issued by the India Meteorological Department before the southwest monsoon commences is a typical example.

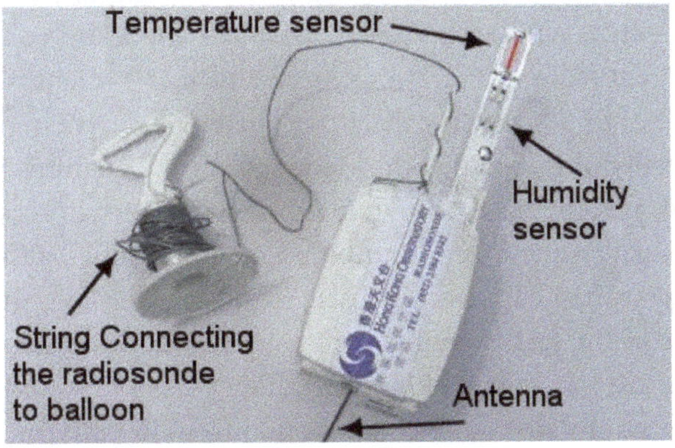

Fig. 1: Forecasting instrument

3.6. PREDICTABILITY

Future weather is governed by the birth, growth and decay of various atmospheric systems, as also by their interaction and transformation in space and time. It is not always not always possible to identify and track all the individual systems for the entire course of their life cycles. This introduces an inherent limitation in the predictability of weather forecasts as the period of validity increases. Thus, shorter the range, the higher is the degree of accuracy of forecasts.

3.7. USES OF FORECASTS

Accurate weather data, forecasts and warnings disseminated through the newspapers and over the radio and enable millions of people to make decisions for effective resource management and safety of life and property. The period of validity and accuracy of weather forecasting have improved with advent of sophisticated observing systems, remote sensing, real time telecommunications, atmospheric modeling and super fast computing capabilities.

Weather simply refers to the condition of air on the earth at a given place and time. It is a continuous, data-intensive, multidimensional, dynamic and chaotic process. These properties make weather forecasting is a formidable challenge. Forecasting is the process of estimation in unknown situations from the historical data. Weather forecasting is one of the most scientifically and technologically challenging problems around the world in the last century. To make an accurate prediction is indeed, one of the major challenges that meteorologists are facing all over the world. Since ancient times, weather prediction has been one of the most interesting and fascinating domains. Scientists have tried to forecast meteorological characteristics using a number of methods, some of these methods being more accurate than others.

Knowledge of meteorology forms the basis of scientific weather forecasting, which revolves around predicting the state of the atmosphere for a given location. Weather forecasting as practiced by humans is an example of having to make judgments in the presence of uncertainty. Weather forecasts are often made by collecting quantitative data about the current state of the atmosphere and using scientific understanding of atmospheric processes to project how the atmosphere will evolve in future. Over the last few years, the necessity of increasing knowledge about the cognitive process in weather forecasting has been recognized. For human practitioners, forecasting the weather becomes a task for which the details can be uniquely personal, although most human forecasters use approaches based on the science of meteorology in common to deal with the challenges of the task. Weather forecasting entails predicting how the present state of the atmosphere will change. Present weather conditions are obtained by ground observations, observations from ships, observation from aircraft, radio sounds, doppler radar and satellites. This information is sent to meteorological centres where the data are collected, analysed and made into a variety of charts, maps and graphs. Modern high-speed computers transfer the many thousands of observations onto surface and upper-air maps. Weather forecasts provide critical information about future weather. There are various techniques involved in weather forecasting, from relatively simple observation of the sky to highly complex computerized mathematical models. Weather prediction could be one day/one week or a few months ahead. The accuracy of weather forecasts however, falls significantly beyond a week. Weather forecasting remains a complex business, due to its chaotic and unpredictable nature.

It remains a process that is neither wholly science nor wholly art. It is known that persons with little or no formal training can develop considerable forecasting skill. For example, farmers often are quite capable of making their own short-term forecasts of those meteorological factors that directly influence their livelihood, and a similar statement can be made about pilots, fishermen, mountain climbers, etc. Weather phenomena, usually of a complex nature, have a direct impact on the safety and/or economic stability of such persons.

Accurate weather forecast models are important to third world countries, where the entire agriculture depends upon weather. It is thus a major concern to identify any trends for weather parameters to deviate from its periodicity, which would disrupt the economy of the country. This fear has been aggravated due to threat by the global warming and greenhouse effect. The impact of extreme weather phenomena on society is growing more and more costly, causing infrastructure damage, injury and the loss of life.

As practiced by the professionally trained meteorologist, weather forecasting today is a highly developed skill that is grounded in scientific principle and method and that makes use of advanced technological tools. The notable improvement in forecast accuracy that has been achieved since 1950 is a direct outgrowth of technological developments, basic and applied research, and the application of new knowledge and methods by weather forecasters. High-speed computers, meteorological satellites, and weather radars are tools that have played major roles in improving weather forecasts. Several other factors have contributed significantly to this increase in forecasting accuracy. One is the development of statistical methods for enhancing the scope and accuracy of model predictions. Another is the improved observational capability afforded by meteorological satellites. A third primary reason for the increase in accuracy is the continued improvement of the initial conditions prepared for the forecast models. Statistical methods allow a wider variety of meteorological elements to be predicted than do the models alone, and they tailor the geographically less precise model forecasts to specific locations. Satellites now provide the capability for nearly continuous viewing and remote sensing of the atmosphere on a global scale. The improvement in initial conditions is the result of an increased number of observations and better use of the observations in computational techniques.

3.8 TRADITIONAL WEATHER FORECASTING

Weather forecasting is the application of science and technology to predict the state of the atmosphere for a given location. Weather forecasts are made by collecting quantitative data about the current state of the atmosphere and using scientific understanding of atmospheric processes to project how the atmosphere will evolve. There are a variety of end users to weather forecasts. Weather warnings are important forecasts because they are used to protect life and property.

In ancient times, forecasting was mostly based on weather pattern observation. Over the years, the study of weather patterns has resulted in various techniques for rainfall forecasting. Present rainfall forecasting embodies a combination of computer models, interpretation, and an acquaintance of weather patterns.

The following technique was used for existing weather prediction.

3.8.1 Use of a Barometer

Measurements of barometric pressure and the pressure tendency have been used in forecasting since the late 19th century. The larger the change in pressure, the larger the change in weather can be expected. If the pressure drop is rapid, a low-pressure system is approaching, and there is a greater chance of rain.

3.8.2 Looking at the sky

Along with pressure tendency, the condition of the sky is one of the most important parameters used to forecast weather in mountainous areas. Thickening of cloud cover or the invasion of a higher cloud deck is an indication of rain in the near future. At night, high thin clouds can lead to halos around the moon, which indicates the approach of a warm front and its associated rain. Morning fog portends fair conditions, as rainy conditions are preceded by wind or clouds which prevent fog formation.

3.8.3 Nowcasting

The forecasting of the weather within the next six hours is often referred to as nowcasting. In this time range, it is possible to forecast smaller features such as individual showers and thunderstorms with reasonable accuracy, as well as other features too small to be resolved by a computer model. A human, given the latest radar, satellite and observational data will be able to make a better analysis of the small-scale features present and so will be able to make a more accurate forecast for the following few hours.

3.8.4 Analog Technique

The analog technique is a complex way of making a forecast, requiring the forecaster to remember a previous weather event which is expected to be mimicked by an upcoming event. It remains a useful method of observing rainfall in places such as oceans, as well as the forecasting of precipitation amounts and distribution in the future. A similar technique is used in medium range forecasting, which is known as teleconnections, when systems in other locations are used to help pin down the location of another system within the surrounding regime.

3.8.5 Numerical Weather Prediction model

Numerical Weather Prediction (NWP) is the science of predicting the weather using models of the atmosphere and computational techniques. Current weather conditions are used at the input of the mathematical models of the atmosphere to predict the weather. This model usually provides surrounding point around the wind farm with a spatial resolution of a few kilometers. NWP uses the power of computers to make a forecast. A forecaster examines how the features predicted by the computer will interact to produce the day's weather. The NWP method is flawed in that the equations used by the models to simulate the atmosphere are not precise.

A number of weather forecasting agencies operate modeling centers where supercomputers are used to run NWP models that span the entire globe. These include the National Center for Environmental Prediction (NCEP) in the United States, the United Kingdom Meteorological Office (UKMO), and the European Centre for Medium-rangeWeather Forecasts (ECMWF). Although costly, a global approach to NWP is essential, especially for long-range forecasting. For this reason, achieving accurate forecasts requires an accurate analysis from which to get the model started. This involves a computer-based process called data assimilation, in which the most recent weather observations from around the world are combined with model forecasts to create a global analysis of current conditions. This becomes the starting point for the next run of the NWP model, and is the computer equivalent of the manual analysis cycle that forecasters carry out on an on-going basis. Global models play a key role in modern weather forecasting, and meteorologists at Met Service routinely use the NCEP, UKMO and ECMWF models to assist with day-to-day production of forecasts and weather warnings. These models give insight into the behavior of weather systems on a large scale, without much emphasis on local detail.

3.8.6 Ensemble Forecasting

To predict the weather forecast meteorologists have developed atmospheric models that approximate the atmosphere by using ensemble forecasting to describe how atmospheric temperature, pressure and moisture will change over time. The equations are programmed into a computer and the data on the present atmospheric conditions are fed into the computer. The computer solves the equations to determine how the different atmospheric variables will change over the next few minutes. The computer repeats this procedure again and again using the output from one cycle as the input for the next cycle.

For some desired time in the future, the computer prints its calculated information. It then analyzes the data, drawing the lines for the projected position of the various pressure systems. A forecaster uses the prognostic chart as a guide to predicting the weather. There are many atmospheric models that represent the atmosphere, with each one interpreting the atmosphere in a slightly different way. Weather forecasts made for 12 and 24 hours are typically accurate. Forecasts made for two or three days are usually good. Beyond above five days, forecast accuracy falls off rapidly. Weather information can also come from remote sensing, particularly radar and satellites.

3.8.6.1 Radar

Radar stands for Radio Detection and Ranging. In radar, a transmitter sends out radio waves. The radio waves bounce off the nearest object and then return to a receiver. Weather radar can sense many characteristics of precipitation, its location, motion, intensity, and the likelihood of future precipitation. Most weather radar is Doppler radar, which can also track how fast the precipitation falls. Radar can outline the structure of a storm and in doing so estimates the possibility that it will produce severe weather condition.

3.8.6.2 Weather satellites

Weather satellites have been increasingly important sources of weather data since the first one was launched in 1952. Weather satellites are the best way to monitor large scale systems, like storms. Satellites can also monitor the spread of ash from a volcanic eruption, smoke from fires, and pollution. They are able to record long-term changes.

Figure 2 shows one of the geostationary satellites that monitors conditions over the world. Weather satellites may observe all energy from all wavelengths in the electromagnetic spectrum. Most important are the visible light and infrared (heat) frequencies.

Fig. 2: Geostationary satellites

3.8.7 Weather Map:

Weather maps simply and graphically depict meteorological conditions in the atmosphere. Weather maps may display only one feature of the atmosphere or multiple features. They can depict information from computer models or from human observations. Weather maps are found in newspapers, on television, and on the Internet. On a weather map, each weather station will have important meteorological conditions plotted. These conditions may include temperature, current weather, dew point, cloud cover, sea level air pressure, wind speed and direction. On a weather map, meteorologists use many different symbols. These symbols give them a quick and easy way to put information onto the map. Figure 3 shows some of these symbols used for Weather Map.

Generally, there are two methods used in weather forecasting namely the empirical and dynamical approach . occurrence of analogues and is often referred to by meteorologists as analogue forecasting. This approach normally is useful for predicting local recorded cases are plentiful. Dynamical approach is based upon equation and forward simulations of the atmosphere and is often referred to as computer modeling. This dynamical approach is useful for modeling large predict short-term weather efficiently. Most of the weather forecasting systems combine both empirical and dynamical approaches.

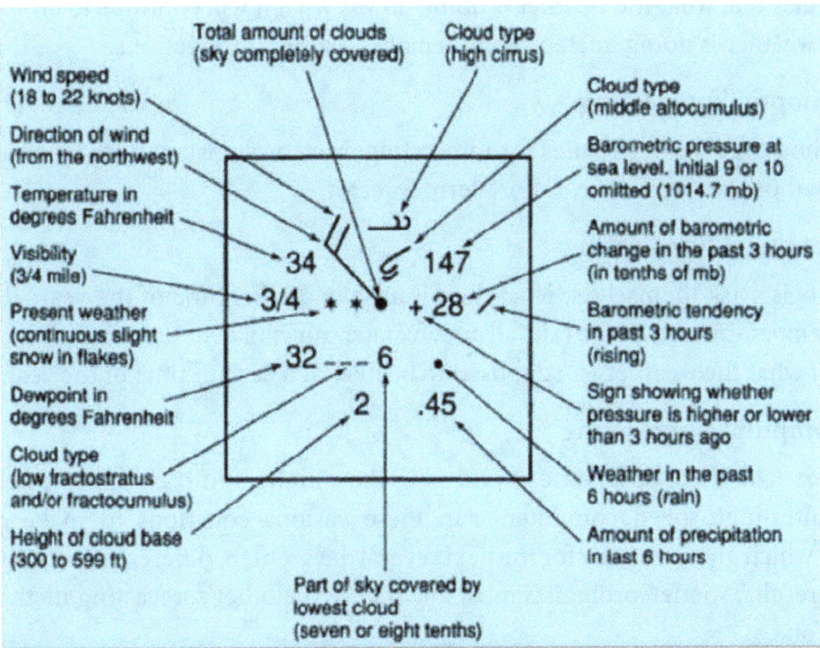

Fig. 3: Weather Map

3.9 TYPES OF WEATHER FORECASTING

A daily weather forecast involves the work of thousands of observers and meteorologists all over the world. Modern computers make forecasts more accurate than ever, and weather satellites orb Figure3 Weather Map The empirical approach is based upon the local-scale weather if large-scale weather phenomena and may not rival orbiting the earth take photographs of clouds from space. Forecasters use the observations from ground and space, along with formulas and rules based on experience of what has happened in the past, and then make their forecast. Meteorologists actually use a combination of several different methods to come up with their daily weather forecasts [HTTP1]. They are:

- Persistence Forecasting
- Synoptic Forecasting
- Statistical Forecasting
- Computer forecasting

3.9.1 Persistence Forecasting

The simplest method of forecasting the weather is persistence forecasting. It relies upon today's conditions to forecast the conditions tomorrow. This can be a valid way of forecasting the weather when it is in a steady state, such as during the summer season in the tropics. This method of forecasting strongly depends upon the presence of a stagnant weather pattern. It can be useful in both short-range forecasts and long-range forecasts.

This assumes that what the weather is doing now is what it will continue to do. To find out what the weather is doing, meteorologists make weather observations.

3.9.2 Synoptic Forecasting

This method uses the basic rules for forecasting. Meteorologists take their observations, and apply those rules to make a short-term forecast.

3.9.3 Statistical Forecasting

Meteorologists ask themselves, what does it usually do this time of the year? Records of average temperatures, average rainfall and average snowfall over the years give forecasters an idea of what the weather is "supposed to be like" at a certain time of the year.

3.9.4 Computer Forecasting

Forecasters take their observations and plug the numbers into complicated equations. Several ultra-high-speed computers run these various equations to make computer "models" which give a forecast for the next several days. Often, different equations produce different results, so meteorologists must always use the other forecasting methods along with this one.

Using all the above methods, forecasters come up with their "best guess" as to what weather conditions will be over the next few days.

Weather forecasting now has a wide range of operational products that traditionally are classified under the following groups:

- Very short-range forecast
- Short-range forecast
- Medium-range forecast
- Long-range forecast

Each weather forecast can be defined on the basis of the following criteria:

- Dominant technology
- Temporal range of validity after emission
- Characteristics of input and output time and space resolution
- Broadcasting needs
- Accuracy

3.10 IMPORTANCE OF WEATHER FORECASTING

Weather forecasting is used in many situations like severe weather alerts and advisories, predicting the behavior of the cloud for air transport, prediction of waterways in a sea, agricultural development and avoiding forest fire.

3.10.1 Severe Weather alerts And Advisories

A major part of modern weather forecasting is the severe weather alerts and advisories which are the national weather service's issue in anticipation of severe or hazardous weather are expected. This is done to protect life and property. Some of the most commonly known of severe weather advisories are the severe thunderstorm and tornado warning, as well as the severe thunderstorm and tornado watch. Other forms of these advisories include winter weather, high wind, flood, tropical cyclone, and fog. Severe weather advisories and alerts are broadcast through the media, including radio, using emergency systems as the Emergency Alert System which breaks into regular programming.

Predicting the behavior of the cloud for Air transport The aviation industry is especially sensitive to the weather and accurate weather forecasting is essential. Fog or exceptionally low ceilings can prevent many aircraft from landing and taking off. Turbulence and icing are also significant in-flight hazards.

Thunderstorms are a problem for all aircrafts because of severe turbulence due to their updrafts and outflow boundaries, icing due to the heavy precipitation, as well as large hail, strong winds, and lightning, all of which can cause severe damage to an aircraft in flight. Volcanic ash is also a significant problem for aviation, as aircraft can lose engine power within ash clouds [JRD06].

3.10.2 Prediction of Waterways in a Sea

Commercial and recreational use of waterways can be limited significantly by wind direction, speed, wave periodicity, high tides and precipitation. These factors can each influence the safety of marine transit. Consequently, a variety of codes have been established to efficiently transmit detailed marine weather forecasts to vessel pilots via radio, for example marine forecast. Typical weather forecasts can be received at sea through the use of Radio fax.

3.10.3 Agricultural Development

Weather plays an important role in agricultural production. It has a profound influence on the growth, development and yields of a crop, incidence of pests and diseases, water needs and fertilizer requirements in terms of differences in nutrient mobilization due to water stresses and timeliness and effectiveness of prophylactic and cultural operations on crops. Weather aberrations may cause (i) physical damage to crops and (ii) soil erosion. The quality of the crop produced during movement from field to storage and transport to market depends on weather. Bad weather may affect the quality of the produce during transport and viability and vigor of seeds and planting material during storage.

3.10.4 Avoiding Forest Fire

Weather forecasting of wind, precipitations and humidity is essential for preventing and controlling wildfires. Different indices, like the Forest fire weather index and the Haines

Index, have been developed to predict the areas more at risk to experience fire from natural or human causes. Conditions for the development of harmful insects can also be predicted by weather forecasting.

3.10.5 Military Applications

Military weather forecasters present weather conditions to the war fighter community. Military weather forecasters provide pre-flight and in-flight weather briefs to pilots and provide real time resource protection services for military installations. Naval forecasters cover the waters and ship weather forecasts. The Navy provides a special service to both themselves and the rest of the federal government by issuing forecasts for tropical cyclone across the Pacific and Indian Oceans through their Joint Typhoon Warning Center.

3.10.6 Air Force

Air Force Weather provides weather forecasting for the Air Force and the Army. Air Force forecasters cover air operations in both wartime and peacetime operations and provide Army support. Military and civilian forecasters actively cooperate in analyzing and creating weather forecast products.

4

METEOROLOGICAL INSTRUMENTS

4.1 METEOROLOGICAL OBSERVATIONS: GENERAL

The meteorological observations of rainfall, air temperature, soil temperature, soil moisture, humidity, evaporation, evapotranspiration, cloudy rains, radiation, sunshine hours, dew, wind speed and direction. The occurrence of phenomenon like thunderstorm, hailstorm, dust storm, frost, cold wind, drought and flood are also weather abnormal features affecting agricultural production. Therefore, the record of meteorological observation of experimental area must include daily observation as well as abnormal weather features to interpret the relationship.

OBSERVATORY: It is a place for recording meteorological observations.

4.2 METEOROLOGICAL OBSERVATORIES

IMD has three types of meteorological observatory spread all over the country with central forecasting and training station at Pune.

These are of three types-

4.2.1 Class 'A' Type

At these stations, continuous record of weather elements is made. These observatories are provided with manual and self-recording instruments.

Its Main Activities

a. Recording large number of meteorological parameters related to agriculture.
b. Preparation of agroclimatic areas and measurements of evaporation.
c. Balloon observation is also recorded at some of these stations (Temperature, pressure, relative humidity).

It is also called as principal agrometeorological observatory.

4.2.2 Class 'B' Type

At these observatories, observations are made twice a day for various weather elements. These are also provided with eye reading and self-recording instruments. It is also called ordinary Agrometeorological observatories.

There are two types of observatories in this category.

A. Crop Weather Observatories

These observatories record side by side meteorological data on growth and yield of one or more major crops (rice, sorghum, wheat, cotton, sugarcane) according to standard plan. Then the data are sent to agromet division, Pune, where these are compiled and weather relationships will be drawn.

B. Non-Crop Weather Observatiories

This class includes-

1. Agrometeorological observatories located at soil conservation research and demonstration centers.
2. Agrometeorological observatories located at agricultural schools, colleges and Universities. The NARP (National Agricultural Research Project) divided the country into 127 agro climatic zones with a regional agricultural research station (RARS) for each agro climatic zone.

3. Class 'C' Type

The C-type observatories or Auxiliary agrometeorological observatories send qualitative data on pest and diseases. These observatories are usually agricultural farms.

4.3 AGRICULTURAL METEOROLOGICAL NETWORK

The agrometeorological stations are divided into 5 categories-

- Ordinary agricultural meteorological stations
- Auxiliary agricultural meteorological stations
- Principal agricultural meteorological stations
- Pilot balloon cum micro meteorological stations
- Evapotranspiration stations

4.4 SOME USEFUL TERMS

- Altitude- Vertical height above sea level.
- Equator- It is the imaginary line lying midway between the poles on the earth surface. It is also the longest circumference of the earth.
- Latitude- It is the distance of a place in north and south of the equator. It runs from west to east and parallel to the equator, at 10 intervals 360 latitudes are drawn.

- Longitude- It is the distance of a place east or west of prime meridian.
- *Indian standard time (IST) is calculated from longitude of Allahabad. Longitude of Allahabad is 82 ½ degree East (82.50E).
- *Prime meridian passes through Greenwich of London.
- Poles- The two points at northern and southern extremities of the earth known as North Pole and South Pole respectively.
- Local time- It is calculated by the calculation of the position of sun. It is also called as solar time. In view of the rotation of the earth from west to east on earth axis, place lying on the same latitude will thus have same noman and same local time. As the earth rotates through 3600 once in 24hrs, local time changes by an hour for every 150 longitudes.

  So,

  150=1hr (60mins)

  150 = 60mins

  So, 10 =60/15=4mins.
- Standard time- The surface of the earth is divided into 24 zones according to the hours in a day, these are called time zones and each zone extends over 150 longitudes. In India longitude is 82.50 East near Allahabad is selected for Indian standard time (IST).
- GMT (Greenwich Mean Time)- It Means time measured from Greenwich situated on 00 Longitude in U.K.

4.5 AGROMETEOROLOGICAL WEATHER ELEMENTS

The main objective is to set up an agricultural meteorological observatory is to take observations of the following weather elements characterizing the agricultural environment influencing the growth and development of different biotic factors-

- Solar radiation and sunshine / cloudiness
- Air and soil temperature
- Relative humidity / dew point
- Wind speed and wind direction
- Soil moisture
- Evaporation (E)
- Evapotranspiration (ET)
- Precipitation (Rain, snow, hail, drizzle)

- Microclimatic Observations
- Continuous record of rainfall, temperature, humidity with self-recording devices.

4.6 SELECTION OF SITE FOR AGROMETEOROLOGICAL OBSERVATORY

Following points must be kept in view-

- The ideal site of an agromet. observatory is well exposed, bare leveled plots having length of 55m in north-south direction and breadth of 36m in east-west direction.
- The agromet observatory should be located more or less at the center of research farm.
- The plot should represent the natural soil topography free from any salts and concretion layers.
- The observatory should well away from high mountains, buildings, trees, high structure, main irrigation channels, rivers, water tanks.
- Site should be free from water lodging area.
- Proper fencing of observatory should be done around site.

The observatory is established with standard meteorological instruments. The important characteristics for selection of instruments are-

- Reliability
- Accuracy
- Sensitivity
- Simplicity of design
- Convenience of operation and maintenance

4.7 AGROMETEOROLOGICAL INSTRUMENTS ARE DIVIDED INTO 4 CATEGORIES

- Conventional (traditional) and ordinary instruments
- Self-recording instruments
- Semi-automatic instruments
- Fully automatic data collection systems

Details of meteorological instruments s. No.	Weather parameter	Name of instruments	Accuracy	Observation frequency	Time of observations
1	Rain fall	a. Ordinary rain gauge b. Self-recording rain gauge	0.2± mm 0.5± mm	Twice daily Continuous (24hrs)	08:30 & 14.30 IST Chart Setting at 08:30 IST
2	Air temperature	a. Maximum thermometer b. Minimum thermometer c. Bimetallic Thermometer d. Grass minimum thermometer	0.5±°C 0.5±°C 0.250±C 0.5±°C	Twice Daily Twice Daily Continuous Once Daily	07:00 & 14:00 IST 07:00 & 14:00 IST Continuous (Chart Setting at 08:30 IST) Just before sunrise
3	Evaporation	USWB Open pan Evaporimeter	0.1± mm	Twice daily	08:30 & 14:30 IST
4	Relative Humidity (R.H)	a. Dry & Wet bulb thermometer b. Assmann Psychrometer c. Hair Hygrograph	0.10±C 0.10±C %2±	Twice daily Twice Daily Continuous	07:00 & 14:00 IST 07:00 & 14:00 IST Chart setting at 08.30 IST
5	Wind speed	Cup anemometer	0.1±Km	Twice Daily	07:00 & 14:00 IST
6	Wind direction	Wind vane	100±	Twice Daily	07:00 & 14:00 IST
7	Sunshine duration	Sunshine recorder	1 meter/ hour	Continuous	Card fitting before sunrise
8	Soil Temperature	Soil thermometer at different depth	±0.10C	Twice daily	07.00 & 14:00 IST
9	Dew	Dew gauge	±0.1 mm	Once daily	Before sunrise
10	Microclimatic data	Assmann Psychrometer	±0.10C	Based on requirement	--------------------

4.8 OBSERVATORY TIMINGS

In agromet observatory some observations are to be recorded at 07:00hrs and 14:00hrs LMT (Local mean time of Allahabad). This is also IST of Allahabad. To find out LMT of different stations we should first find the difference between the longitude of a particular station with respect to Allahabad longitude as,

LMT=IST±4 ×(82030/E-Station longitude)

Where, /=Minute

The timing of observation for rainfall and evaporation is at 08:30 & 14:30 IST.

4.9 GENERAL PRECAUTIONS FOR INSTALLATION OF METEOROLOGICAL INSTRUMENTS

1. All the tall instruments should be kept on north side of the observatory so that their shadow may not fall on the other instruments.
2. The face of the Stevenson screen is always open in the north direction in northern hemisphere, so that the direct solar radiation may not fall on the instrument while recording the observation.
3. All the instruments should be installed according to the norms and instruction of IMD.
4. Soil thermometer should always face south direction in northern hemisphere.
5. The instrument should be properly well spaced.

4.10 MEASUREMENT OF RAINFALL

Rainfall is measured in term of depth of rain water (in mm) on a solid surface if rain water neither allowed to percolate nor runoff nor lost in any form. The device used for measurement of rainfall is called rain gauge.

There are two types of rain gauges-

1. Ordinary or non-recording type rain gauge.
2. Recording or Self-recording type rain gauge.

4.10.1 ORDINARY RAIN GAUGE-

The instrument is designed to measure the rainfall amount. It is simple and easy instrument to operate. The standard rain gauge is Simon's pattern widely accepted all over the world.

It consists of a collector with a gun metal rim funnel, a base and a bottle. The collector has a deep-set funnel and the complete rain gauge has a slight tapper with the narrower portion at the top. The rain gauge has the capacity of 200mm of rainfall.

The amount of rainfall collected by the gauge is measured with help of a measuring cylinder known as rain measure. Separate rain measures are available for 100cm2 and 200cm2 collector rain gauges. For a 100cm2 collector (diameter-112.9mm) 100cc of water is equal to 10mm of rain fall. For 200 sq cm collector, 200 cc of water is equal to 10mm of rain.

Installation

The rain gauge is placed on a leveled ground at the standard height of 30cm above the ground surface. The structural work is constructed with the standard principles.

Precautions

- The rain water should be poured carefully into the rain measure taking care to avoid spilling of any drop of water.
- The rain gauge should be placed on a horizontal surface while recording rainfall.

4.10.2 Self Recording Rain Gauge

There are three types-

- Tipping Bucket type
- Weighing Type
- Floating Type

This instrument is designed for giving continuous recording of rainfall with time. It can be used to find out total amount of rainfall, time of onset and closing of a rainy event and intensity of rainfall.

4.11 PRINCIPLE

Rain water entering the gauge at the receiving funnel is carried through a filter to a receiver consisting of a float and syphon arrangement. The pen is mounted on the stem of float. As the water level rises in the receiver the float rises and record the rain on the graph fixed on a clock mounted drum. The clock drum records once in 24hrs, so that continuous record of rainfall curve is obtained. The vertical scale of the graph records the amount of rainfall and after 10 mm of accumulated rainfall the syphonic action take place with draining of the water. The pen will be come to the bottom of the graph. It again rises by receive of rainfall. If there is no rain the pen draws a horizontal line. The slope of rainfall curve gives the intensity of the rainfall with time.

4.12 INSTRUMENT DESCRIPTION-

The main parts are-

- Float Chamber
- Funnel
- Strip of bottom of float
- Entrance tube
- Float
- Float rod
- Clock drum
- Syphon chamber and tube
- Chart no-395 is used for harvesting 10mm capacity of rainfall scale on vertical side.

4.13 OPERATION

The chart of the rain gauge is daily changed at 08.30 IST. The pen should be set at 0 reading after pouring sufficient water into the receiver till the pen reaches the top of the chart and water is drained out by syphonic action. The pen should be cleaned and use fresh ink every day for good marking on chart paper.

4.14 THERMOGRAPH

It is an automatic instrument with particular record of temperature.

There are two types of the thermograph-

a. Bourdon tube type
b. Bimetallic type

Bimetallic thermograph consists of sensitive elements of two strips of different metal. i.e., Brass and iron welded together along their flat surfaces and bent into arc, one end of the arc is fixed to the base of instrument and other end is connected to the pen which draws changes on graph paper. Changes in temperature cause two metal to expand or to contrast by different amount so that they bend and unbend and pen traces accordingly. This instrument should be placed in double Stevenson screen. It should be standardized twice a day.

4.15 UNIT AND DIMENSION

UNIT- The unit is a physical quantity defined as the reference standard used to measure it.

Physical Quantity- These are length, mass, time and quantity of a substance.

Need for unit- Measurement is the process of comparison. For the measurement of a physical quantity we need to have the parameters.

$$Q=nu$$

Where,

Q= Quantity

n= Number

u= Unit

From the equation, we can draw some conclusion that small size of unit means more will be the value of n for a given physical quantity.

4.15.1 Characteristics of a Unit

A unit should have following basic characteristics-

- Unit should be well defined
- Of suitable size

- It should be coherent
- It should be simple in use

4.15.2 Category of Unit

Unit are divided into following categories- Fundamental Unit	Supplementary Unit	Derived Units
Unit in which fundamental Physical Quantity (Length, Mass, time, quantity of substance, temperature) are expressed.	Units in which plane and solid angle are expressed.	Units in which derived quantities like speed, pressure, energy etc. are expressed are called derived units.

4.16 BASIC PRINCIPLE OF SCIENTIFIC INSTRUMENTS

Basic principle involves in all scientific instruments is related to the physical phenomena to also related to the sense of users by same process i.e., thermometer measures changes in the degree of warmth of atmospheric air. In this case change in warmth is a physical phenomenon, quantity involve is degree of warmth (temp), expansion of mercury is a physical process which correlates the change in the warmth to the sense of user.

4.17 CHARACTERISTICS OF A GOOD INSTRUMENT

1. An instrument should be accurate as required for the problem in hand. Ex-If temperature variation within ±0.50C, a thermometer graduated in 0.50 will be sufficient for accuracy.

The following are the basic requirements for the accuracy of instruments

- All instruments should have and maintain calibration under a given condition within the required precision. The error under other condition should be known and should be constant in time within required limit.
- The instrument should be sensitive. Sensitivity of the instruments refers to the change in final message to be sensitive by the user per unit change in quantity to be measured.

*Sensitivity and accuracy are different and not be confused i.e., more sensitive instrument with incorrect scale is less accurate.

- The instrument should be reliable means there should be minimum or no chance of failing the instrument during the recording period.
- It should be compact and durable and should have minimum handling problem.
- It should be of low cost and easy maintenance.
- An instrument should respond immediate to change in the quantity being measured.

4.18 MICROMETEOROLOGICAL INSTRUMENTS

Thermometer-The device used to measure temperature is called as thermometer.

Meteorological thermometer are classified as under- s. no.	Class	Basic principle	Example
1	Liquid in glass Thermometer	Liquid expands on heating	Mercury thermometer
2	Deformation Thermometer	Solid expands on heating	Bimetallic thermometer
3	Liquid in metal Thermometer	Liquid expands on heating	Mercury in steel & other combination
4		Electrical Thermometers	
a.	Resistance thermometer	Resistance of a conductor increases with increase in temperature.	Platinum resistance thermometer
b.	Thermistor	Resistance of a semiconductor decreases with increase in temperature.	Platinum resistance thermometer
c.	Thermo couple Thermometer	See back effect	
5	Radiation thermometer	Stephan Boltzmann law of radiation	Infrared thermometer

4.19 ELECTRICAL THERMOMETERS

4.19.1 Resistance Thermometer

Resistance thermometer is based on the fact that resistance of a conductor increases with increase in temperature and increase is given by-

$$R2=R1\ [1+\alpha\ (t2-t1)]$$

Where, R1 and R2 are resistances at temperature t1 and t2 respectively.

$\alpha\alpha$=Temperature coefficient of a material of thermometer.

Platinum, nickel and copper are the metal used for constructing a resistance thermometer. Of these platinum resistance thermometer is preferred because of the following reasons-

- It has large temperature coefficient.
- It is resistance to corrosion.
- It is able to maintain its characteristics over a longer period.

Source of error

- Heat produced by current passing through the sensor charge the resistance hence recorded value is more than actual value.
- Needs calibration time to time.

4.19.2 Thermistor

It is a temperature measuring device based on a fact that resistance of certain material like germanium and silicon decreases with increase in temperature. Such material has negative temperature coefficient and their resistance is related to temperature as under.

$$Rt=R0ae-b/t$$

Where, a & b are constant.

Rt=Resistance at temperature t.

R0 =Resistance at 00temperature.

Advantage-

- It is more sensitive as change in resistance during temperature is larger.
- It has low or less time lag as sensor size is very small.
- It needs calibration check.

4.19.3 Thermocouple Thermometer

These are based on thermoelectric/see back effect. According to this effect, when two wires of different thermal properties (physical) are joined together and one end of the junction heated keeping other cold, the emf is produced.

The pair of wire so joined is called thermocouple. The emf produced depends on following factor-

- Nature of material of which a thermocouple is made (bismuth antimony material). Thermocouple is most sensitive.
- The temperature difference between hot and cold junction i.e.,

$$V= \alpha\alpha\theta+1/2\beta\beta\theta2$$

Where,

V=emf

$\beta\beta,\alpha$ are Constant.

θ=Temperature difference between both junctions.

Advantages-

- Calibration curve is linear up to a range of temperature and measurement needed in micrometeorological studies.
- Lag time is very low as sensor size is a point size.

Demerits

- Low output needs sensitive milli microvolt meters.
- It needs calibration check from time to time.

4.19.4 RADIATION THERMOMETER-

Infrared thermometer is one of such kind of thermometer which sense the surface temperature without making any physical contact with the surface. It works on the principle of Stefan Boltzmann's law of black body radiation.

According to this law radiation emitted by a surface at any temperature is given by-

$$E=\sigma\sigma\sqrt{T}T4$$

E=Energy radiated per unit time per unit area (w/m2)

σ=Stefan constant (5.67×10-8 w/m2/k4)

T=Temperature of the surface (0K)

Advantage

It does not disturb the surface whose temperature is to be measured.

4.20 RADIOMETERS-

Radiometers are used to measure solar radiation. The sensor used in various radiation thermometer is a thermopile which is a series combination of a number of thermocouples.

Radiometers are classified according to their use as given below- sl. No	Radiometers	Use	Sensor/principle
1	Albedometer	To measure reflected solar radiation from crop and other surfaces.	Thermopile (See back effect)
2	Pyrheliometer & Solarimeter	To measure global diffuse and Direct radiation.	-Do-
3	Net radio meter	Use to measure overall net radiation.	Do
4	Soil heat flux plate	Use to measure heat conducted inside the soil	Do
5	Quantum sensor	Use to measure PAR	Photodiode

*All the radio meters mostly in use except quantum sensor used thermopile sensor.

Structure of Radiometer

The blackened surface of the sensor consists of alternate thin strip of manganin and constantan with one set of the junction along the central line of surface while the remaining junction are in good thermal contact with the relatively massive supporting poles which are insulated electrically not thermally from the base plate (reference junction). The temperature difference between the active reference junctions of sensor system sets emf which is a function of solar radiation and can be recorded on suitable multiple meter.

The Surface receiving radiation is covered with hemispherical glass zone. It protects the sensor from wind and rain and also prevents from the convective currents. It helps in bringing blackened surface into thermal equilibrium that is loss of heat by all causes is equal to the gain of heat due to the radiation. This state is essential for thermo emf to the function of radiation only.

The whole instrument is supported in a solid brass case, outer part of which is highly polished to prevent direct absorption of radiation so that temperature of the case kept near the temperature of the surrounding air. The air bubble is provided on the base of instrument to make the surface horizontal.

4.21 QUANTUM SENSOR

This sensor is sensitive within photosynthetically active radiation (PAR) with spectral distribution of wave length ranging between 400 to 700 nm.

4.22 PSYCHROMETER

It is the device used to measure actual vapour pressure and hence the state of dampness of atmosphere. A psychrometer consists of two thermometers placed in air side by side, one is known as dry bulb thermometer and second is wet bulb thermometer.

1. Dry bulb thermometer
2. Wet bulb thermometer

*Both thermometers are Hg in glass thermometers.

Hence both thermometers are similar. Wet bulb is wrapped in a muslin piece of cloth. In the month of May-June (hot summer) wet bulb depression is more as compared to rainy season (Because of more evaporation and high temperature wet bulb have depression).

The cloth is kept wet by supply of water from reservoir through capillary action. As the air passes by the wet bulb, water from wet surface absorbs it from its surrounding air and evaporates (causes cooling). So, in this process (Td to Tw), air surrounding the wet bulb cools from temperature. The basic principle of psychrometer is a fact that temperature difference between dry and wet bulb is a function of actual vapour pressure (AVP). If temperature difference is more then vapour pressure deficit (VPD) is more.

There are mainly two types of psychrometer-

- Unaspirated psychrometer
- Aspirated psychrometer

*The difference between dry bulb temperature (Td) and wet bulb temperature (Tw) is called as wet bulb depression.

4.22.1 Unaspirated Psychrometer

It is a simple psychrometer consisting of a set dry and wet bulb installed in a screen with no artificial ventilation. From the dry and wet bulb reading the vapour pressure is calculated using the following formula.

i. When Tw <60°C

$$e_a = e_{sw} - \frac{0.48(Td-Tw)P}{670-Tw}$$

ii. When Tw>40°C

$$e_a = e_{sw} - \frac{0.48(Td-Tw)P}{610-Tw}$$

Where,

e_a = Actual vapour pressure

e_{sw} = Saturated vapour pressure

P= atmospheric pressure

4.22.2 Aspirated Psychrometer

In this type of psychrometer the air is drawn over the thermometer artificially to maintain the state of equilibrium.

5

WEATHER HAZARDS AND ITS IMPACT

5.1 INTRODUCTION

- A weather hazard is an extreme weather event that threatens people or property. Weather hazards include: tropical storms. tornadoes droughts.
- Extreme weather events, and climatic anomalies, have major impacts on agriculture. Of the total annual crop losses in world agriculture, many are due to direct weather and climatic effects such as drought, flash floods, untimely rains, frost, hail, and storms

Atmospheric processes and energy exchanges are driven by Earth's energy balance and linked to climate and weather. Hurricanes, thunderstorms, tornadoes, blizzards, ice storms, dust storms, heat waves, as well as flash flooding resulting from intense precipitation, are all natural processes that are hazardous to people. These severe hazards affect considerable portions of Northern America and are responsible for causing significant death and destruction each year.,

- Weather is the measure of short-term conditions of the atmosphere such as temperature, wind speed, and humidity. The average weather of a region over decades is its climate. Although weather can vary day-to-day or year-to-year, the climate of a region is relatively stable because it represents the average weather over a long period of time.
- Extra-warm summers and extra-cold winters, when combined with typical seasonal change, result in a moderate average temperature over long periods of decades or centuries. Proximity to a large body of water can also decrease the temperature range of a geographic region.

5.2 TYPES OF WEATHER HAZARDS

5.2.1. Frost and High Temperature

Frost:

- Frost occurs when the air temperature near the ground surface falls below 0°C. Low temperatures generally retard the growth of the crop plants. Cold advection is more injurious during winter season because it creates a typical hazard when the field crops are in their seedling stage.
- However, wheat crop can withstand freezing temperature but the plants are killed, if the roots are disturbed by frost heave. Frost occurrence is more common in middle and high latitudes. It occurs rarely in the tropical regions. However, it may occur on the high mountains in the tropical regions. The quality of the mature crops can be reduced by sub- freezing temperatures.
- Frost is very injurious to the vegetable crops. Crops in north-west India experience frost of moderate to severe intensity. Therefore, every year potato and tomato crops are damaged worth crores of rupees due to low temperature injuries.
- These crops are susceptible to frost injury right up to the maturity. The flowering stage is a critical period for most of the field crops and young fruit plants suffer serious damage due to low temperature injuries.
- To protect the plants from the frost, there are many techniques in which the temperature is not allowed to fall below frost level:
    - The temperature of the plant is increased by increasing the soil temperature. The soil temperature can be increased by giving irrigation to the crop.
    - By covering the plants with glass or plastic covers. In this way, the temperature of the plants rises.
    - Sprinkling irrigation also increases the temperature of the air.

High Temperature:

- High temperatures are experienced during summer season in many parts of the tropical and sub-tropical regions. Prolonged high temperature conditions can lead to heat wave conditions. Generally heat wave conditions can occur during the months of April, May and even June.
- If the maximum temperature remains above normal by 6-7°C, moderate heat wave conditions are said to occur. If the maximum temperature remains above normal by 8°C or more, then it may be termed as severe heat wave.
- Heat waves often develop over Rajasthan, Haryana and Punjab areas, which are far away from the coastal areas. At the same time, hot and strong north-westerly winds may cause heat waves over the coastal areas of Orissa and Andhra Pradesh.

However, heat wave incidence is rare over the peninsula south of the latitude 13° north because of maritime influence and fairly humid conditions.

- Generally, heat wave conditions may persist for 4-5 days. But sometimes they persist for another one week. The intensity of heat wave is found to be maximum during the months of May and June. Heat wave generally extends from north-west India right up to Odisha and Andhra coast. Every year highest severity of heat waves is found in Uttar Pradesh and Bihar states.
- Hot winds are experienced by many areas in the world. Generally, these winds are generated in those areas which are located under the influence of anticyclonic circulation. Their severity is further enhanced by the advection of hot air from other hot areas.
- These hot winds are also caused when the air mass descends down the slopes. Foehn is one of the hot winds, which is generated by adiabatic heating when the air mass descends on the leeward side of the mountains.
- These hot winds are generally experienced on the northern side of the Alps Mountain in Switzerland. These winds are comparatively warmer and drier than the prevailing air mass of that area. These warm and dry winds can melt the snow rapidly.
- As a result, sufficient water becomes available to the crops which are grown under rainfed conditions. On the other hand, the arrival of these hot and dry winds can increase the air temperature suddenly, resulting adverse effect on the standing crops.
- Similarly, hot and dry winds are also called Chinook winds. These winds prevail on the eastern slopes of the Rocky Mountains. Under favourable conditions, these winds can develop over any mountain range. These winds carry huge thermal energy which can melt the snow very quickly. These winds are also called snow eater.
- Under the influence of these winds, the air temperature can increase by about 22°C within 24 hours. On 27th January 1940, a rise in temperature of 14°C was recorded in two hours at Danver, Colorada. These warm and dry winds can reduce the severity of the winter in the western parts of the great plains in north America.
- High temperatures are generally found over the land in the tropical areas. During summer season, if the maximum and minimum temperature remain above normal for a few days, then excessive thermal energy accumulates which decreases the relative humidity drastically. Under such conditions, the water requirements of the crop plants increase manifold resulting adverse impact on the growth of the plants.

- Under high temperature conditions, the growth of the plants is retarded. The vegetable crops are more sensitive to high temperature conditions. The plants can be protected by giving frequent irrigations. Shelterbelts can be raised to decrease the effect of high temperature.

Global Transfer of Heat

If the planet is trying to balance the extremes of the poles and equator, how is heat transferred around the planet? There are four ways: *conduction*, *convection*, *latent heat*, and *radiation*.

Conduction: Conduction is the transfer of heat from a warmer object to a cooler object through molecule interaction. As the Sun heats the ground, energy is transferred to the atmosphere by conduction. But the atmosphere is a very poor conductor of heat. In calm weather, the heated ground only warms the first few centimeters of air. In fact, the atmosphere's temperature can be 50 percent colder at your head than near your feet. Since the atmosphere is such a poor conductor of heat, there must be other ways to transfer the energy.

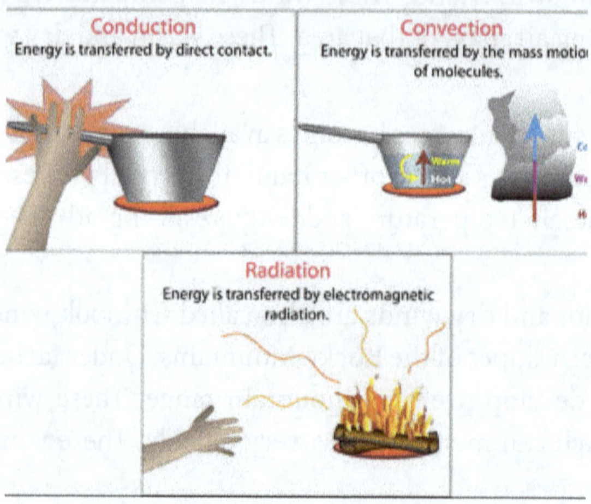

Fig. 1: Conduction, Convection and Radiation

Convection: Convection is the transfer of heat by the mass movement of a fluid (such as water and air). It occurs mostly in liquids and gases because they are free to move around. Heat is transferred upward and outward away from its heat source and cooler air is brought in to replace the rising air. On a local scale convection in the summer can produce afternoon thunderstorms, but on a global scale convection transfers energy from the equator toward the poles and back.

Radiation: The final process of transferring heat around the planet is called radiation. The energy coming from the Sun comes in various wavelengths. A wavelength is the distance measured along a wave of energy from one crest to another; the shorter the wavelength, the higher the energy. The energy received from the Sun passes through the atmosphere without warming it. But all objects that absorb radiation from the sun radiate some of that energy back out in a weaker form of energy called long wave radiation such as heat. All living objects (including humans) radiate longwave radiation. The Earth itself radiates the energy absorbed from the Sun in the form of longwave radiation which is sometimes called Earth light. If you were an astronaut wearing infrared goggles, you would actually see the earth glowing from the radiating infrared radiation.

Latent Heat: Latent heat, also called a phase change, is a powerful force in weather. When water transforms from gas to liquid or solid (or vice versa), it is called a phase change. The heat required to change phases is called latent heat. In order for water to change from a liquid to a gas, energy/heat must be taken from some external source such as the surrounding atmosphere. Therefore, evaporation is a cooling process because the water is taking heat from the surrounding air in order to evaporate. With the image above, visualize that every time evaporation or condensation occurs, there is a direct transfer of heat and energy.

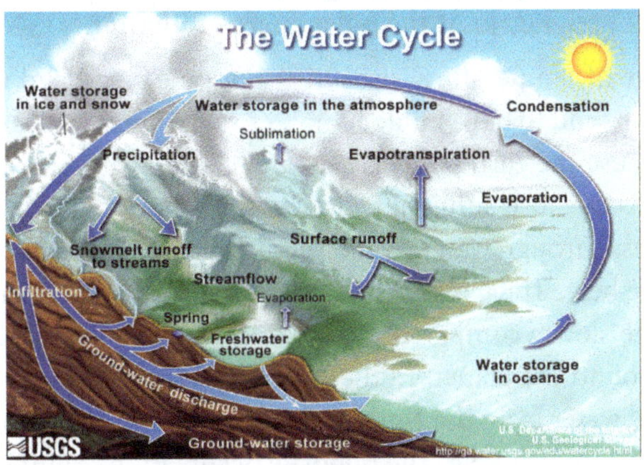

Fig. 2: The water cycle

The process of evaporation and condensation transfers large amounts of heat around the planet. The heat released in a typical thunderstorm is equal to an atomic bomb. The heat/energy within a hurricane is equal to 100,000 atomic bombs and could power the United States for over a year!

Fig. 3: The heat/energy within a hurricane

There are several controlling factors that determine global temperatures. The first and most significant is latitude. Because of the Earth's shape the angle of the Sun hitting the planet, temperatures are greatest near the equator and decrease toward the poles. In fact, at the equator more energy is absorbed from the Sun than is radiated back into space. At the poles, more energy is radiated back into space than is absorbed by the Sun. The purpose of weather and ocean currents is to balance out these two extremes.

Fig. 4: At the poles energy radiation

The next control of temperature is the distribution of land and water on the planet. Places near the ocean tend to have milder climates year-round versus regions surrounded by land. This is because the earth can heat up and cool down faster and greater than the ocean. The reason is because sunlight must heat a larger volume of area in the ocean because light can pass through water. Water requires 5 times more energy to heat up than earth, called specific heat. Thus, region's temperatures located near large bodies of water temperatures change slowly compared to land. Ocean currents are also vital controls in transferring heat around the planet. In the Northern Hemisphere, ocean currents rotate clockwise bring cold water from the North Pole toward the equator and warm water from the equator toward the North Pole. The opposite occurs in the Southern Hemisphere.

Fig. 5: ocean currents rotate clockwise bring cold water from the North Pole toward the equator

The last control of temperature is elevation. On average, the air decreases in temperature with altitude about 3.6 degrees Fahrenheit per 1000 feet, called the normal lapse rate or temperature lapse rate. Using the Wasatch Front in Utah as an example, Example in Utah: elevation in Salt Lake City is about 6,500 feet above sea level and the Wasatch Mountains elevations are about 10,000 feet above sea level. That's 3,500 feet rise in elevation. If it cools at 3.6 degrees Fahrenheit per 1,000 feet, then the top of the mountains is about 12.6 degrees Fahrenheit cooler.

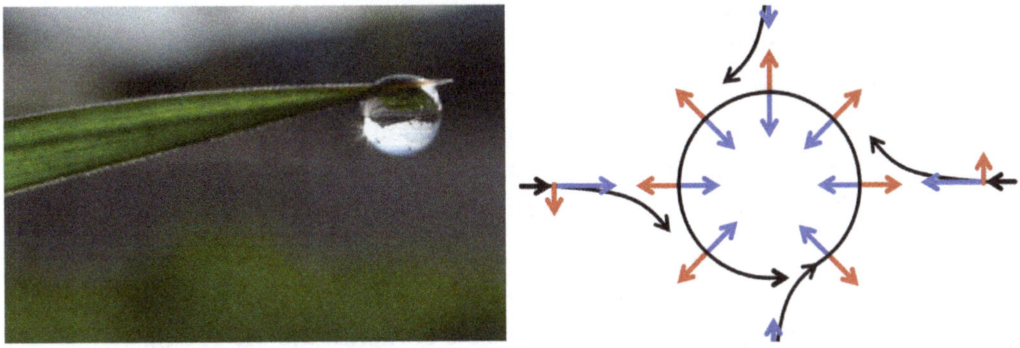

Fig. 6: The normal lapse rate or temperature lapse rate

5.2.2 Drought

- Drought occurs in those areas of the world, where the soil moisture is not sufficient to meet the demands of the potential evapotranspiration. Low relative humidity, wind and high temperatures are the contributory factors, which can create drought conditions by increasing the evapotranspiration.
- It is a common phenomenon in the desert areas, where evapotranspiration exceeds the rainfall. Under such conditions, agriculture is not possible without irrigation.
- Drought is one of the worst disastrous of various natural hazards. Drought is generally considered to be a period of moisture deficiency. Drought occurs whenever the supply of moisture from rainfall or stored in the soil becomes insufficient to fulfil the optimum water needs of the plants.
- Drought is such a phenomena, whose effect is felt after it has happened. Under prolonged drought conditions, it is not possible to raise agricultural crops. Therefore, drought conditions impose a great threat to the agricultural production.
- The water requirements of the plants vary from season to season and from place to place. The water requirement of a crop depends upon the weather conditions prevailing during different stages of the crop.
- At the same time, the stage of the crop is very important, therefore the water requirement increases from early stage to the reproductive stage of the crop. Inadequate soil moisture availability during reproductive stage has detrimental effects on the yield.

- Drought creates adverse impact on the agriculture. There may be failure of crops under severe drought conditions. Thus, the prolonged drought conditions can shatter the economy of a region.
- Drought can be divided into four types:
  - Permanent droughts,
  - Seasonal droughts,
  - Contingent droughts, and
  - Invisible droughts.

Permanent Droughts: Permanent droughts are found in the desert area, where rainfall is not equal to the water needs of the plants. In such cases, evaporation always exceeds total rainfall during the life cycle of the crops. Agriculture is not possible without irrigation.

Seasonal Droughts: Seasonal droughts occur in those areas, where there are well-defined rainy and dry seasons. These droughts are expected every year. Agriculture is possible during the rainy season and it is possible only with the use of irrigation in dry season.

Contingent Droughts: Contingent droughts occur when rainfall is erratic and variable. These droughts are found in sub-humid and humid areas. These droughts can occur in any season but these are more severe during the periods of greatest water need. They are serious because they cannot be predicted. The crops stand wilt under the influence of these three droughts.

Invisible Droughts: These droughts cannot be recognised very easily. Invisible drought can occur at any time. It may occur even during rainy season when the daily supply of moisture fails to meet the daily water requirements of the plants. Invisible droughts are very harmful for the crops. Yield of the crop is adversely affected under these conditions. High yield can be obtained by supplying irrigation to the crop.

5.2. 3 Floods

- Floods are the major weather hazards, which are caused by heavy rainfall over a given area in a short period of time. In some of the areas, flood producing storms follow seasonal pattern, whereas in other areas flood producing storms occur irregularly.
- These floods cause heavy damage to the crops and agricultural buildings. Northern and Eastern parts of India are prone to floods, where agricultural crops are adversely affected. The floods cause a greater loss of life and property than any other natural disaster.
- There are three types of floods:
  - ***First type:*** First type of flood is caused due to intense local rainfall. In such cases, a very large amount of water is precipitated over a small area within a

short time. Under these circumstances, the rate of arrival of water at the surface of the earth is much greater than the infiltration rate into the saturated soil. In such cases, flash floods can occur.

The flash floods are most common in those areas which experience heavy thunderstorms. Therefore, these may be regarded as a potential hazard whenever intense rainfall occurs. In the arid climates, the thunder showers resulting flash floods are erratic. This type of flash floods are common during monsoon season in the hilly areas.

- ***Second type:*** Second type of flood occurs, when snow begins to melt rapidly. This happens only when the rise of temperature is associated with rain. Warm rain melts snow more quickly than hot sunshine, most of which is reflected by the white surface. In mountainous areas of mid and high latitudes, snow melt results flooding with the onset of the warm season. Such floods have severe agricultural effects.
- ***Third type:*** Third type can be called autumn or winter flood, which is caused by rainfall lasting for many days. Although the rate of rainfall may be quite less but the total precipitation over a period of one or more days may be considerable. The fast moving water droplets can damage the delicate crop plants and seriously affect their growth and hence yield is reduced.

5.2.4 Storms

(a) Tropical Storms/Thunder Storms

- Tropical storms/Thunder storms are the most destructive weather phenomenon. Many parts of the earth experience these storms. Several thousands thunder storms occur every day mainly in the tropics. Their number is smaller over the oceans than over the land, owing to lack of intense convection above water surface. Thunder storms rarely occur in the polar areas.
- These are always associated with unstable air and strong vertical motions, that produce cumulonimbus clouds. They derive their energy from the release of latent heat of condensation in the rising humid air.
- The tropical storms develop in those areas of the oceans, where temperature at the sea surface exceeds 26°C. High temperature conditions lead to the formation of a low pressure area. As a result, the winds get organised in the form of cyclonic circulation. A low pressure area gets intensified into a depression due to the availability of abundant water vapours.
- Later on, pressure decreases rapidly and depression changes into a cyclone. The tropical storms occur in Bay of Bengal and Arabian Sea. They cause lot of damage to the agricultural crops and in some cases the tropical storms bring needed rains to the drought affected agricultural area.

Conditions favourable for the development of thunder storms

- Strong convection due to intense heating of the land surface.
- Passage of cold, moist air mass over the warm water surface.
- Forced ascent of conditionally unstable air along the convergence zone or along the slope of the mountains.
- Radiational cooling at upper levels.
- Cold advection aloft and warm advection at the surface.

Generally thunder storms caused by the surface heating over land are most common during summer afternoon and early evening. But thunderstorms occur over the ocean during night time because water surface is warmer than the air aloft.

(b) Hail Storms

- Hailstorms are the worst weather hazards. Generally hail storms develop in the cumulonimbus clouds. Large hails are always associated with thunderstorms. Hails are generally confined to a small area of the storm. Large hailstones cause a great damage to the life and agricultural crops. In India hailstorms occur during winter season.
- Their intensity increases during the months of March and April when the wheat crop reaches the harvesting stage. Thus crops worth crores of rupees are damaged every year.

(c) Dust Storms

- Dust storms generally occur during summer season when the atmospheric pressure decreases suddenly. The wind speed can reach up to 100 km/hr and in some cases the speed may even exceed 100 km/hr. Trees are severely affected by the dust storms and the electric poles are uprooted.
- Normally dust storms do not cause rainfall because sufficient water vapours are not available for the formation of the clouds. However, if the amount of water vapours increases and becomes sufficient for the formation of the clouds, thunder showers may occur.

5.2.5 Tornadoes and Water Spouts

- The tornado is generally associated with cumulonimbus clouds. It extends downward from the base of the cloud in the form of a chimney. This chimney has both translatory as well as rotatory motion. Therefore, it has the tendency to touch the land surface.
- The pressure within the chimney falls rapidly and may be exceptionally low at the centre of the chimney as compared to the adjoining atmospheric air outside the chimney.

- As a result, a tremendous force is generated within the chimney. Wherever it touches the ground, the force within the chimney can suck or lift the big objects above the ground surface causing great damage to the objects coming on its way.
- The diameter of the chimney vary from 10 meter to 100 meter or even 1km to 2km. The damage caused by the tornado is further enhanced by the strong surface winds associated with the storm.
- Tornadoes occur in many areas of the world. Maximum number of tornadoes is found in U.S.A. The word tornado refers to the most violent storm in which the speed of the surface wind can exceed 400 km/hr over a small area. The maximum wind speed in a tornado has never been measured. The vertical velocity can exceed 250 km/hr. These tornadoes cause lot of damage to life and property and agricultural buildings.
- There has been no universal theory of the tornado formation. Generally collision of two different air masses may lead to the development of a tornado. Instability occurs, when a dry, cold (heavy) polar air mass pushes the warm, humid (light) air mass upwards. The rising warm, humid air mass loses temperature at the dry adiabatic lapse rate.
- The air mass becomes cold and saturated, which leads to condensation. Huge amount of latent heat released during condensation keeps the rising air mass warm. This causes the air currents reach greater heights inducing very low pressure in the center of the air column, which becomes a site of strong winds.
- As a result, cyclonic circulation along with strong updraft leads to the formation of a funnel shaped storm. As a result, cyclonic circulation along with strong updraft leads to the formation of a funnel shaped storm cloud associated with extremely loud roaring noise and unusually intense lightening.

Water spouts: When a tornado occurs on the water surface / surface of the ocean, it is called water spout. Whenever the chimney of the tornado touches the water surface, it sucks water upwards and sometimes it may lift small vessels in the oceans. But the damage caused by the water spout is less as compared to the tornado over the land.

5.2.6. Lightening:

- Lightening is also a natural calamity and it is always associated with a cumulonimbus cloud. The vertical extent of this cloud may extend up to 10-16 km height above the ground surface. Such clouds consist of positive charge or even potential difference is created between the cloud and the earth.
- As a result, a giant sparkle of lightening is generated. Sometimes a lightening flash may hit any object like a tree or a building on the ground.
- In the United States, lightening accounts for more deaths than any other weather hazard. The lightening casualties generally occur in the afternoon. Lightening

can strike the aircraft causing heavy damage. It also causes a substantial amount of property damage. Forest fire generally occurs because of lightening under dry conditions.

5.2.7 Blizzards

- Blizzards are another weather hazard which cause heavy damage to the life and property. The combination of very low temperatures and very strong winds and snow storm is called blizzard.
- These are generally found in high latitudes where extra tropical cyclones are prevalent during winter season.

5.2.8 Earthquakes and Tsunami Waves

- Of all the natural calamities, earth quakes and tsunami waves are the worst calamities of the world, which cause great damage and devastation in those areas where they occur.
- India has experienced some of the world's most devastating earthquakes in the recent years.

5.3 AGRICULTURE AND WEATHER RELATIONS

Weather includes many factors such as sunshine, temperature, precipitation, evaporation, transpiration etc. which is depicted as under:

- ***Solar Radiation:*** It includes light intensity, light quality and duration of sunlight. Out of total radiation received, only 50% in photsynthetically active radiation(PAR) which lies in 400-700nm range. Rest is UV or IR. Now there is an exponential relation between amount of light intercepted by canopy and Leaf Area Index (Leaf Area/ground area). The sum of these values for individual days is directly proportional to crop yield.
- ***Temperature***: A term growing degree days is used which is given for a crop from time of flowering to harvest date. Every crop has a temperature range below or above which GDD for that particular day is zero. GDD is the sum of difference of daily temperature and base temperature. So the harvest date is predicted depending on the when the GDD is achieved.
- ***Precipitation, Evaporation and Transpiration:*** There is a relation to calculate the length of growing period which includes Evapotranspiration. You need to know amount of precipitation and crop water requirement. Also transpiration causes cooling so maintains temperature. This prevents crop from damaging.
- These factors are responsible for proper crop growth and development. If anyone of the factors is affected, the crop will be affected accordingly as in case of heavy rains, high temperatures etc.

6

ENVIRONMENTAL PHYSICS

6.1 COMPOSITION AND DIVISIONS OF THE ATMOSPHERE

The earth has three great realms: Lithosphere (solids), Hydrosphere (water) and Atmosphere (gases). The gases realm extends over the earth's surface to a height of about 9,600km and envelops it. It consists of several gases and minute suspended liquid and solid particles.

6.1.1 Major Constituents

Air is a Mixture of several gases. Four gases: Nitrogen, Oxygen, Argon & carbon dioxide account for more than 99 percent of the dry air.

6.1.2 Constituent By Volume (%)

Nitrogen 78.088

Oxygen 20.949

Argon 0.93

Carbon Dioxide 0.033

Total 100.00

These gases are mixed in constant proportions up to 80 km. In addition to these gases, very minor amount of neon, helium, krypton, xenon, methane, hydrogen, nitrous oxide, carbon monoxide, ozone, ammonia, nitrogen oxide, sulphur dioxide comprise the remaining part. Along with these gases, water vapour which is more variable in its occurrence in time and space is vital atmospheric constituent.

6.1.3 Divisions of the Atmosphere

On the basis of composition, the atmosphere is broadly divided into two spheres. Homosphere, the zone of homogenous composition of gases and Heterosphere, the zone of variable composition. Homosphere extends to about 80km above the earth surface and the composition of the atmosphere is almost uniform at different levels.

Homosphere is further subdivided in to 4 layers.

1. Troposphere
2. Stratosphere
3. Mesosphere
4. Thermosphere

The transition zone between two spheres is called pause such as-

- Tropopause (Transition zone between troposphere and stratosphere)
- Stratopause (Transition zone between stratosphere and Mesosphere)
- Mesopause (Transition zone between mesosphere and Thermosphere)

Troposphere

- This atmospheric layer is nearest layer to the earth's surface, extending to around 16 to 17 km. Extension of this layer is not same at all latitudes. It is about 17 km at the equator, 12.5km at 450 latitude and 4.5 km over polar region.
- This layer contains almost all the water and dust and represents the belt of all atmospheric phenomena-clouds, rain, snow, fog, lightning, thunderstorms, cyclone etc. No visible weather phenomenon occurs above this layer.
- In this layer, the temperature decreases at 6.50C/km, which is known as lapse rate.
- In the troposphere, instead of decrease in temperature with altitude, sudden abrupt raise in temperature is seen at certain points, which is known as temperature inversion. For instance, the lapse rate changes abruptly at 14 km.
- Convection current is present and wind speed increases with height.

Stratosphere

- The stratosphere is about 50 km thick at equator and about 70 km at the poles.
- Temperature remains nearly constant (isothermal) in lower layers and increases very gradually up to stratosphere. Lower stratosphere within 20 km is more isothermal than the upper stratosphere.
- There are no convection currents. Cirrus clouds may occasionally form in the lower atmosphere.

Mesosphere

- It extends up to about 80 km from the earth's surface. Maximum chemical activity takes place in this sphere and hence it is also called chemosphere.

- It is a warmer layer due to selective absorption of ultraviolet rays by ozone. Temperature increases with height at about 5 ^{0}C/km and reaches a maximum around 77 ^{0}C at about 40-50 km. Then the temperature falls to as low as around -100 ^{0}C at the top of mesosphere.
- It contains ions which reflects ratio waves back to earth surface and enables us to have wireless communication.

Thermosphere

- It extends upward from 80 km.
- Temperature increases rapidly (around 1000 ^{0}C at 350 km) with height due to absorption of ultraviolet radiation.
- Its lower portion consists of atomic oxygen and nitrogen.

Above the thermosphere is ionosphere where ionization of atmosphere begins and temperature increases with height due to very low air density. According to some scientists, thermosphere includes ionosphere.

6.2 HEAT TRANSFER

Heat energy moves from a hotter object to a cooler one. The movement of heat energy will stop when the temperature of each object becomes the same. The three methods by which heat moves are called conduction, convection and radiation.

One end of the metal rod is being heated by the Bunsen burner. After a period of time, the other end of the rod becomes heated. Heat energy has travelled along the rod. This method of heat transfer is called conduction. The only group of substances that can be called good conductors of heat are metals. All other substances are classed as insulators (bad conductor of heat).

Transfer of heat by movement of a mass or substance from one place to another, generally vertical is called convection. The third way in which heat energy is transferred is called radiation or more properly heat radiation. The heat energy is carried by waves called electromagnetic waves. The type of electromagnetic wave which is felt as heat is called 'infrared'. These heat waves can travel through a vacuum and this is one of the forms of energy that reaches us from the sun.

6.2.1 Heating of Atmosphere

Sun is the ultimate source of atmosphere heat and energy, but its effect is not direct for example as we climb a mountain or ascend toward the sun in an airplane, temperature become steadily lower rather than higher as we might expect. This is because the mechanism of heating the atmosphere are not simple. There are three heating process directly responsible for heating the atmosphere.

They are:

6.2.2 Radiation

When the source of heat, transmit heat directly to an object through heat waves, it is known as radiation process. In this process heat travels through the empty space. The vast amount of heat energy coming to and leaving the earth is in the form of radiation.

- All objects whether hot or cold, emit radiant energy continuously.
- Hotter objects radiate more energy per unit area than colder objects.
- Insolation reaches the earth's surface in short waves and heat is radiated from the earth in long waves.

The atmosphere is transparent to short waves and opaque to long waves, hence energy leaving the earth's surface heats up the atmosphere more than the solar radiation.

6.2.3 Conduction

The atmospheric conduction occurs at the zone of contact between the atmosphere and the earth's surface. However, this is a minor method of heat transfer in terms of warming the atmosphere. Since it only affects the air close to the earth. This is because air is a bad conductor of heat unlike certain metals.

6.2.4 Convection

The air of the lower layers of the atmosphere gets heated either by earth's radiation or by conduction. The heating of the air leads to its expansion. Its density decreases ant it moves upwards. Continuous ascent of heated air creates vacuum in the lower layer of the atmosphere. As a consequence, cooler air comes down to fill the vacuum leading to convection. The cyclic movement associated with the convectional process in the atmosphere transfer heat from the layer to the upper layer and heats up the atmosphere.

As the beam of sunlight enters the atmosphere, it first passes through the mesosphere with little change. In the stratosphere, the density of atmosphere gases increases. There is more oxygen available, which reacts with the shortest or ultra violet wavelengths and effectively removes them. The atmosphere, is thus warmed in the process. By now about 2 percent of the sun's rays has been lost.

Then in troposphere, it loses about 48% of sun's rays due scattering in many directions by the gas molecules, dust particles etc.

By the time, the sun rays reach the ground surface they retain with only 50 percent of its original energy. Even then not all of this is absorbed. This is because the surface itself has an albedo. The albedo varies from surface to surface. For example, the albedo of freshly fallen snow may reach as high as 90%. The greatest variable is over water. When the sun is high in the sky, water has a very low albedo. This is why the oceans appear dark on satellite photographs at low angles of the sun, such as at dusk at dawn or in midwinter in sub polar latitude and temperate latitudes, the albedo may reach nearly 80 percent.

HEAT LOST BY conducts no heat), but each being in contact of a third system C through a diathermal wall which permit heat to pass. Then the system A will be in thermal equilibrium with system C and similarly system B will also be in thermal equilibrium with system C.

Reflection/scattering	Heat lost by absorption
1.Reflected by clouds=24%	1.Absorbed by clouds=3%
2.Reflected /Scattered by air + dust+ water=6%	2.Absorbed by air + dust + water vapour=14%
3.Total reflection by atmosphere=30%	3.Total absorption by atmosphere=17%
4.Reflection from earth surface=6%	4.Absorbed by earth surface=47%
Total reflection (Earth + Atmosphere) =36%	Total Absorption (Earth + Atmosphere) =64%

It gives general conclusion:

If two systems are in thermal equilibrium with a third system then they must be in thermal equilibrium with each other.

This statement is called the zeroth law of thermodynamics.

Mechanical Equilibrium

For a system to be in mechanical equilibrium there should be no unbalanced forces existing between different parts of the system or between the system and the surroundings.

Thermal Equilibrium

For a system to be in thermal equilibrium, the temperatures at all parts of the system must be the same and should be identical with that of the surroundings.

Chemical Equilibrium

For a system to be in chemical equilibrium, the composition of the system should remain fixed and definite.

A system which satisfies all the above three equilibriums i.e., uniformity of pressure, temperature and chemical composition is said to be the thermodynamic equilibrium.

6.3 THERMODYNAMIC PROCESS

6.3.1 Isothermal Process

If a thermodynamic system is perfectly conducting to the surroundings and undergoes a physical process in such a way that its temperature remains constant throughout the process is said to be isothermal process.

The equation connecting the pressure P and Volume V of one mole of a gas for an isothermal process is

PV=RT=CONSTANT

For n Moles of a gas-

$$PV= nRT$$

In isothermal changes, there is no change in temperature, no change in internal energy i.e., dU =0

$$dQ =0+dW$$

$$\text{Or, } dQ=dW$$

Thus, during an isothermal process, Heat added (or subtracted) = work done by (or on) the gas.

6.3.2 Isobaric Process

If the working substance is taken in an expanding chamber in which the pressure is kept constant, the process is called isobaric process.

6.3.3 Isochoric Process

If a substance undergoes a process in which the volume remains unchanged the process is called an isochoric process.

6.3.4 Adiabatic Process

If a thermodynamic system is perfectly insulated from the surroundings and undergoes a process in such a way that no exchange of heat takes place between it and the surroundings the process is said to be adiabatic process. In such a process, no heat is allowed either to enter the system or to leave it, but all along the process there is a change in temperature.

6.3.5 Reversible Process

If a process can be carried out in such a way that the effects produced by on the system as well as on the surroundings can be completely restored to their initial states and no changes are left in any of the system taking part in the process or in the surroundings then the process is said to be reversible process.

6.3.6 Irreversible Process

If a process does not satisfy these conditions, it is called irreversible. Thus, the process which produce a permanent change in the thermodynamic state of the system and cannot be retraced in the opposite order are known as irreversible process.

6.4 THE FIRST LAW

The first law of thermodynamics may be stated in two slightly different forms.

1. The first form of this law is one which express the equivalence between mechanical work and heat energy. According to it "when mechanical work is spent in producing heat, a definite quantity of heat is produced for each unit of work spent and conversely, when heat is employed to do work, the same

definite quantity of heat disappears for every unit of work obtained". Thus if W is the amount of work, then the amount of heat Q, that can be produced by it is given by $W \propto Q$

$$W=JQ$$

Where,

$$J = 4.2 \text{ X } 107\text{ergs/cal}$$

$$= 4.2 \text{ J/cal}$$

$$= 4.2 \text{ x } 103\text{Joule/kilo cal.}$$

This form of law is true only if the whole of work done is used in producing heat or vice versa.

2. The second statement of this law is a particular form of the general law of the conversation of energy. Let us suppose that a quantity dQ of heat is supplied to a body. It is in general, spent in 3 ways-
   a. Partially, it is spent in raising the temperature of the body. This is equivalent of increasing its internal kinetic energy.
   b. A part of it is spent in increasing the internal potential energy of the body.
   c. A third part of it is spent in doing external work.

So, dQ=dU +dW.................... eqn (1)

This equation represents the differential form of first law of thermodynamics, which may therefore be stated "In all transformations, the energy due to heat supplied must be balanced by the external work done plus the increase in internal energy".

While using equation (1) two things should be kept in mind,

1. dQ, dU and dW are all to be measured in the same units i.e., all the three either in joules or in calories or kilo calories.
2. If heat is taken by the system then dQ is +ve and if it is given by the system then dQ is –ve. Similarity if the work is done by the system then dW is +ve and if the work is done by some external agency on the system then dW is –ve.

6.4.1 Significance of the 1st Law

The first law of thermodynamics establishes an exact relationship between heat and work. According to it a definite quantity of heat will produce a definite amount of work and vice versa. It denies that work or energy can be created out of nothing. It means that it is impossible to construct a thermal machine which may operate without any expenditure of fuel and may thus create energy out of nowhere. A machine that would do this could run itself and is generally called a perpetual motion machine of the first kind.

Three ideas are included in first law of thermodynamics;

- Heat is a form of energy in transit.
- Conservation of energy takes place in thermodynamics system.
- Every system in equilibrium, possess an internal energy which is a function of state.

6.4.2 Specific Heat of Gases

Suppose that an amount of heat Q is given to a body. If its temperature rises by dT then the heat capacity of the body is defined as:

Heat capacity = Q/dT

Heat capacity per unit mass is known as specific heat and represented by the letter C.

$$\text{Thus, } C = \frac{\text{Heat Capacity}}{\text{Mass}} = Q/m.dT$$

6.4.3 Mass

Thus, specific heat of a material is the quantity of heat required so as to raise the temperature of unit mass of the material through 10C.

In order to find the value of the specific heat of a gas, either the pressure or the volume has to be kept constant. Consequently, we have two specific heats.

a. The specific heat at constant volume

It is defined as the amount of heat required to raise the temperature of unit mass of the gas through 10C when its volume is kept constant. It is denoted by Cv and given by

$$Cv = dQ/dT \times m$$

$$\text{So, } Cv = \frac{dQ}{dT}$$

As m =1

b. The specific heat at constant pressure

It is defined as the amount of heat required to raise the temperature of unit mass of the gas through 10C when its pressure is kept constant. It is denoted by Cp and given by Cp= dQ/dT at constant pressure.

6.5 SECOND LAW OF THERMODYNAMICS (HEAT ENGINES AND ENTROPY)

Heat Engine

Any device which converts heat continuously into mechanical work is called a heat engine.

Thus, for instance, when water is boiled in a vessel closed by a lid, the steam generated inside throws the lid off showing there that high pressure steam can be made to do work.

For any heat engine, there are three essential requirements.

1. Source

A hot body at a fixed temperature T1 from which the heat engine can draw heat is called as source.

2. Sink

A cold body at a fixed lower temperature T2 to which any amount of heat can be rejected is called a sink.

3. Working Substance

The material which on being supplied with heat performs mechanical work is called as working substance.

Thus, in a heat engine the working substance takes in heat from the source converts a part of it into external work, gives out the rest to the sink and returns to its initial state. This series of operation constitute a cycle. The work can be continuously obtained by performing the same cycle over and over again.

Let Q_1 be the amount of heat absorbed by the working substance from the source, Q_2 that rejected by it to the sink and W the net amount of work done by it. Therefore, the net amount of heat absorbed by the working substance is Q_1-Q_2. Since working substance returns to its initial state so change in internal energy, du=0. By first law of thermodynamics-

$$Q_1-Q_2=W \text{.................... (1)}$$

The thermal efficiency of the engine n is defined as the ratio of the net work obtained in the cycle (output) to the heat absorbed by the working substance from the source (input).

So, n= Work output/ Heat input

= W/Q_1

From Equation (1) W = Q_1-Q_2, So n=Q_1-Q_2/Q_1

or n = 1-Q_2/Q_1

or n = 1-T_2/T_1 (in terms of temperature)

If Q_2=0 or in other words if an engine could be built to operate in such a way that no heat at all is rejected by the working substance in a cycle, there will be 100% conversion of heat into work which is not practically possible. (Efficiency of 100%)

Carnot Engine

It is the type of engine which converts all the heat into mechanical work. It is an ideal engine and not practically possible.

6.5 RADIATION LAWS

6.5.1 Plank's Law

Electromagnetic radiation consists of the flow of quanta or particles. The energy content (E) of each quantum is proportional to the frequency and it is given by the following equation.

E=hν

Where,

H = Plank's constant

$=6.625 \times 10^{-27}$erg sec^{-1}

The equation indicates that the greater the frequency the greater is the energy of the quantum.

6.5.2 Kirchoff's Law

Any grey object (other than a perfect black body) that receives radiation disposes of a part of it in reflection and transmission. The values of absorptivity, reflectivity and transmissivity are less than or equal to unity.

This law states that the absorptivity 'a' of an object for radiation of a specific wavelength is equal to its emissivity 'e' for the same wave length.

$$a(\lambda)=e(\lambda)$$

6.5.3 Stefan –Boltzman Law

This law states that the intensity of radiation emitted by a radiating body is proportional to the fourth power of the absolute temperature of that body.

Flux= σT4

Here σ is the Stefan Boltzmann's constant (5.67 x 10-5erg.cm-2sec-1k-4) and T is the absolute temperature of the body.

6.4.4 Wein's Law

The wave length of maximum intensity of emission from a black body is inversely proportional to the absolute temperature (T) of the body. Thus,

Wave length (λ) of maximum intensity (μm)=2897/T

6.4.5 Lambert's Law

This law states that the permeability of the atmosphere to solar radiation. The intensity of solar radiation on a vertical irradiation at the earth's surface is gives by the equation.

$$Im=I_0qm$$

7

MEASUREMENT OF SOIL MOISTURE

7.1 INTRODUCTION

Soil moisture is an important component in the atmospheric water cycle, both on a small agricultural scale and in large-scale modelling of land/atmosphere interaction. Vegetation and crops always depend more on the moisture available at root level than on precipitation occurrence. Water budgeting for irrigation planning, as well as the actual scheduling of irrigation action, requires local soil moisture information. Knowledge of the degree of soil wetness helps to understand the initiation of convective events and to forecast the risk of flash floods or the occurrence of fog. Nevertheless, soil moisture has been seldom observed routinely at meteorological stations. Documentation of soil wetness was usually restricted to the description of the "state of the ground".

7.1.1 Definitions

Soil moisture determinations measure either the soil water content or the soil water potential.

7.1.2 Soil Water Content

An expression of the mass or volume of water in the soil, while the soil water potential is an expression of the soil water energy status. The relation between content and potential is not universal and depends on the characteristics of the local soil, such as soil density and soil texture.

Soil water content on the basis of mass is expressed in the gravimetric soil moisture content, θ_g, defined by:

$$\theta_g = M_{water} / M_{soil} \tag{1}$$

Where M_{water} is the mass of the water in the soil sample and M_{soil} is the mass of dry soil that is contained in the sample. Values of θ_g in meteorology are usually expressed in %.

Because precipitation, evapotranspiration and solute transport variables are commonly expressed in terms of flux, volumetric expressions for water content are often more useful. The volumetric soil moisture content of a soil sample, θ_v, is defined as:

$$\theta_v = V_{water}/V_{sample} \quad (2)$$

Where V_{water} is the volume of water in the soil sample and V_{sample} is the total volume of dry soil + air + water in the sample. Again, the ratio is usually expressed in %, although many research communities are now adopting volumetric water content (m^3/m^3) as the standard for expressing soil moisture. The relationship between gravimetric and volumetric moisture contents is:

$$\theta_v = \theta_g(\rho_b/\rho_w) \quad (3)$$

Where ρ_b is the dry soil bulk density and ρ_w is the soil water density.

The basic technique for measuring soil water content is the gravimetric method, described below . Because this method is based on direct measurements, it is the standard with which all other methods are compared. Unfortunately, gravimetric sampling is destructive, rendering repeat measurements on the same soil sample impossible. Because of the difficulties of accurately measuring dry soil and water volumes, volumetric water contents are not usually determined directly.

7.1.3 Soil water potential.

This describes the energy status of the soil water and is an important parameter for water transport analysis, water storage estimates and soil plant water relationships. A difference in water potential between two soil locations indicates a tendency for water flow, from high to low potential. When the soil is drying, the water potential becomes more negative and the work that must be done to extract water from the soil increases. This makes water uptake by plants more difficult, so the water potential in the plant drops, resulting in plant stress and, eventually, severe wilting.

Formally, the water potential is a measure of the ability of soil water to perform work, or, in the case of negative potential, the work required to remove the water from the soil. The total water potential ψ_t, the combined effect of all force fields, is given by:

$$\psi_t = \psi_z + \psi_m + \psi_o + \psi_p \quad (4)$$

Where ψ_z is the gravitational potential based on elevation above the MSL; ψ_m is the matric potential, suction due to attraction of water by the soil matrix; ψ_o is the osmotic potential, due to energy effects of solutes in water; and ψ_p is the pressure potential, the hydrostatic pressure below a water surface.

The potentials which are not related to the composition of water or soil are together called hydraulic potential, ψ_h. In saturated soil, this is expressed as $\psi_h = \psi_z + \psi_p$, while in unsaturated soil, it is expressed as $\psi_h = \psi_z + \psi_m$. When the phrase "water potential" is used in studies, sometimes with the notation ψ_w, it is advisable to check the author's definition because this term has been used for $\psi_m + \psi_z$ as well as for $\psi_m + \psi_o$.

The gradients of the separate potentials will not always be significantly effective in inducing flow. For example, ψ_o requires a semipermeable membrane to induce flow, and ψ_p will exist in saturated or ponded conditions, but most practical applications are in unsaturated soil.

7.1.4 Units

In solving the mass balance or continuity equations for water, it must be remembered that the components of water content parameters are not dimensionless. Gravimetric water content is the weight of soil water contained in a unit weight of soil (kg water/kg dry soil). Likewise, volumetric water content is a volume fraction (m^3 water/m^3 soil).

The basic unit for expressing water potential is energy (in joules = kg m^2 s^{-2}) per unit mass, J kg^{-1}. Alternatively, energy per unit volume (J m^{-3}) is equivalent to pressure, expressed in pascals (Pa = kg m^{-1} s^{-2}). Units encountered in older literature are bar (= 100 kPa), atmosphere (= 101.32 kPa), or pounds per square inch (= 6.895 kPa). A third class of units are those of pressure head in centimetres of water or mercury, energy per unit weight. The relation of the three potential unit classes is:

$$\psi_t = \psi_z + \psi_m + \psi_o + \psi_p \qquad (5)$$

where γ = 103 kg m^{-3} (density of water) and g = 9.81 m s^{-2} (gravity acceleration). Because the soil water potential has a large range, it is often expressed logarithmically, usually in pressure head of water. A common unit for this is called pF, and is equal to the base 10 logarithm of the absolute value of the head of water expressed in centimetres.

7.1.5 Meteorological Requirements

Soil consists of individual particles and aggregates of mineral and organic materials, separated by spaces or pores which are occupied by water and air. The relative amount of pore space decreases with increasing soil grain size (intuitively one would expect the opposite). The movement of liquid water through soil depends upon the size, shape and generally the geometry of the pore spaces.

If a large quantity of water is added to a block of otherwise "dry" soil, some of it will drain away rapidly by the effects of gravity through any relatively large cracks and channels. The remainder will tend to displace some of the air in the spaces between particles, the larger pore spaces first. Broadly speaking, a well-defined "wetting front" will move downwards into the soil, leaving an increasingly thick layer retaining all the moisture it can hold against gravity. That soil layer is then said to be at "field capacity", a state that for most soils occurs about $\psi_m \approx -33$ J/kg, with a range of values from –1 J/kg for organic soils to –100 J/kg for heavy clay soils. A value of –10 J/kg (pF ≈ 2) can be assigned for a loamy sand soil. This state must not be confused with the undesirable situation of "saturated" soil, where all the pore spaces are occupied by water. After a saturation event, such as heavy rain, the soil usually needs at least 24 h to reach field capacity. When moisture content falls below field capacity, the subsequent limited movement of water in the soil is partly liquid, partly in the vapour phase by distillation (related to temperature gradients in the soil), and sometimes by transport in plant roots.

Plant roots within the block will extract liquid water from the water films around the soil particles with which they are in contact. The rate at which this extraction is possible depends on the soil moisture potential. A point is reached at which the forces holding moisture films to soil particles cannot be overcome by root suction, and plants are starved of water and lose turgidity: soil moisture has reached the "wilting point", which in most cases occurs at a soil water potential of –1.5 MPa (pF = 4.2). In agriculture, the soil water available to plants is commonly taken to be the quantity between field capacity and the wilting point, and this varies highly between soils: in sandy soils it may be less than 10 volume per cent, while in soils with much organic matter it can be over 40 volume per cent.

Usually it is desirable to know the soil moisture content and potential as a function of depth. Evapotranspiration models concern mostly a shallow depth (tens of centimetres); agricultural applications need moisture information at root depth (order of a metre); and atmospheric general circulation models incorporate a number of layers down to a few metres. For hydrological and water-balance needs – such as catchment-scale runoff models, as well as for effects upon soil properties such as soil mechanical strength, thermal conductivity and diffusivity – information on deep soil water content is needed. The accuracy needed in water content determinations and the spatial and temporal resolution required vary by application. An often-occurring problem is the inhomogeneity of many soils, meaning that a single observation location cannot provide absolute knowledge of the regional soil moisture, but only relative knowledge of its change.

7.2 MEASUREMENT METHODS

The methods and instruments available to evaluate soil water status may be classified in three ways. First, a distinction is made between the determination of water content and the determination of water potential. Second, a so-called direct method requires the availability of sizeable representative terrain from which large numbers of soil samples can be taken for destructive evaluation in the laboratory. Indirect methods use an instrument placed in the soil to measure some soil property related to soil moisture. Third, methods can be ranged according to operational applicability, taking into account the regular labour involved, the degree of dependence on laboratory availability, the complexity of the operation and the reliability of the result. Moreover, the preliminary costs of acquiring instrumentation must be compared with the subsequent costs of local routine observation and data processing.

There are five operational alternatives for the determination of soil water content. First, there is classic gravimetric moisture determination, which is a simple direct method. Second, there is lysimetry, a non-destructive variant of gravimetric measurement. A container filled with soil is weighed either occasionally or continuously to indicate changes in total mass in the container, which may in part or totally be due to changes in soil moisture. Third, water content may be determined indirectly by various radiological techniques, such as neutron scattering and gamma absorption. Fourth, water content can be derived from the dielectric properties of soil, for example, by using time-domain reflectometry. Lastly, soil moisture can be inferred on a global scale from remotely sensed measurements of the Earth's thermal or reflective properties.

Soil water potential measurement can be performed by several indirect methods, in particular using tensiometers, resistance blocks and soil psychrometers. None of these instruments is effective at this time over the full range of possible water potential values.

7.2.1 Gravimetric Direct Measurement of Soil Water Content

The gravimetric soil moisture content θ_g is typically determined directly. Soil samples of about 50 g are removed from the field with the best available tools (shovels, spiral hand augers, bucket augers, perhaps power-driven coring tubes), disturbing the sample soil structure as little as possible (Dirksen, 1999). The soil sample should be placed immediately in a leak-proof, seamless, pre-weighed and identified container. As the samples will be placed in an oven, the container should be able to withstand high temperatures without melting or losing significant mass. The most common soil containers are aluminium cans, but non-metallic containers should be used if the samples are to be dried in microwave ovens in the laboratory. If soil samples are to be transported for a considerable distance, tape should be used to seal the container to avoid moisture loss by evaporation.

The samples and container are weighed in the laboratory both before and after drying, the difference being the mass of water originally in the sample. The drying procedure consists in placing the open container in an electrically heated oven at 105 °C until the mass stabilizes at a constant value. The drying times required usually vary between 16 and 24 h. Note that drying at 105 ± 5 °C is part of the usually accepted definition of "soil water content", originating from the aim to measure only the content of "free" water which is not bound to the soil matrix.

If the soil samples contain considerable amounts of organic matter, excessive oxidation may occur at 105 °C and some organic matter will be lost from the sample. Although the specific temperature at which excessive oxidation occurs is difficult to specify, lowering the oven temperature from 105 °C to 70 °C seems to be sufficient to avoid significant loss of organic matter, but this can lead to water content values that are too low. Oven temperatures and drying times should be checked and reported.

Microwave oven drying for the determination of gravimetric water contents may also be used effectively. In this method, soil water temperature is quickly raised to boiling point, then remains constant for a period due to the consumption of heat in vaporizing water. However, the temperature rapidly rises as soon as the energy absorbed by the soil water exceeds the energy needed for vaporizing the water. Caution should be used with this method, as temperatures can become high enough to melt plastic containers if stones are present in the soil sample.

Gravimetric soil water contents of air-dry (25 °C) mineral soil are often less than 2%, but, as the soil approaches saturation, the water content may increase to values between 25% and 60%, depending on soil type. Volumetric soil water content, θ_v, may range from less than 10% for air-dry soil to between 40% and 50% for mineral soils approaching saturation. Soil θ_v determination requires measurement of soil density, for example, by coating a soil clod with paraffin and weighing it in air and water, or some other method.

Water contents for stony or gravelly soils can be grossly misleading. When rocks occupy an appreciable volume of the soil, they modify direct measurement of soil mass, without making a similar contribution to the soil porosity. For example, gravimetric water content may be 10% for a soil sample with a bulk density of 2 000 kg m^{-3} however, the water content of the same sample based on finer soil material (stones and gravel excluded) would be 20% if the bulk density of fine soil material was 1 620 kg m^{-3}.

Although the gravimetric water content for the finer soil fraction, $\theta_{g,fines}$, is the value usually used for spatial and temporal comparison, there may also be a need to determine the volumetric water content for a gravelly soil. The latter value may be important in calculating the volume of water in a root zone. The relationship between the gravimetric water content of the fine soil material and the bulk volumetric water content is given by:

$$\theta_{v,stony} = \theta_{g,fines} \left(\rho_b / \rho_w \right) \left(1 + M_{stones} / M_{fines} \right) \quad (6)$$

Where $\theta_{v,\,stony}$ is the bulk volumetric water content of soil containing stones or gravel and M_{stones} and M_{fines} are the masses of the stone and fine soil fractions.

7.2.2 Soil Water Content: Indirect Methods

The capacity of soil to retain water is a function of soil texture and structure. When removing a soil sample, the soil being evaluated is disturbed, so its water-holding capacity is altered. Indirect methods of measuring soil water are helpful as they allow information to be collected at the same location for many observations without disturbing the soil water system. Moreover, most indirect methods determine the volumetric soil water content without any need for soil density determination.

7.2.2.1 Radiological Methods

Two different radiological methods are available for measuring soil water content. One is the widely used neutron scatter method, which is based on the interaction of high-energy (fast) neutrons and the nuclei of hydrogen atoms in the soil. The other method measures the attenuation of gamma rays as they pass through soil. Both methods use portable equipment for multiple measurements at permanent observation sites and require careful calibration, preferably with the soil in which the equipment is to be used. When using any radiation-emitting device, some precautions are necessary. The manufacturer will provide a shield that must be used at all times. The only time the probe leaves the shield is when it is lowered into the soil access tube. When the guidelines and regulations regarding radiation hazards stipulated by the manufacturers and health authorities are followed, there is no need to fear exposure to excessive radiation levels, regardless of the frequency of use. Nevertheless, whatever the type of radiation-emitting device used, the operator should wear some type of film badge that will enable personal exposure levels to be evaluated and recorded on a monthly basis.

Neutron scattering method: In neutron soil moisture detection, a probe containing a radioactive source emitting high-energy (fast) neutrons and a counter of slow neutrons is lowered into the ground. The hydrogen nuclei, having about the same mass as other nuclei in the soil. Because in any soil most hydrogen is in water molecules, the density of slow "thermalized" neutrons in the vicinity of the neutron probe is nearly proportional to the volumetric soil water content.

Some fraction of the slowed neutrons, after a number of collisions, will again reach the probe and its counter. When the soil water content is large, not many neutrons are able to travel far before being thermalized and ineffective, and then 95% of the counted returning neutrons come from a relatively small soil volume. In wet soil, the "radius of influence" may be only 15 cm, while in dry soil that radius may increase to 50 cm. Therefore, the measured soil volume varies with water content, and thin layers cannot be resolved. This method is hence less suitable to localize water-content discontinuities, and it cannot be used effectively in the top 20 cm of soil on account of the soil air discontinuity.

Several source and detector arrangements are possible in a neutron probe, but it is best to have a probe with a double detector and a central source, typically in a cylindrical container. Such an arrangement allows for a nearly spherical zone of influence and leads to a more linear relation of neutron count to soil water content.

A cable is used to attach a neutron probe to the main instrument electronics, so that the probe can be lowered into a previously installed access tube. The access tube should be seamless and thick enough (at least 1.25 mm) to be rigid, but not so thick that the access tube itself slows neutrons down significantly. The access tube must be made of non-corrosive material, such as stainless steel, aluminium or plastic, although polyvinylchloride should be avoided as it absorbs slow neutrons. Usually, a straight tube with a diameter of 5 cm is sufficient for the probe to be lowered into the tube without a risk of jamming. Care should be taken in installing the access tube to ensure that no air voids exist between the tube and the soil matrix. At least 10 cm of the tube should extend above the soil surface, in order to allow the box containing the electronics to be mounted on top of the access tube. All access tubes should be fitted with a removable cap to keep rainwater from entering the tubes.

In order to enhance experimental reproducibility, the soil water content is not derived directly from the number of slow neutrons detected, but rather from a count ratio (CR), given by:

$$CR = C_{soil}/C_{background} \qquad (7)$$

where C_{soil} is the count of thermalized neutrons detected in the soil and $C_{background}$ is the count of thermalized neutrons in a reference medium. All neutron probe instruments now come with a reference standard for these background calibrations, usually against water. The standard in which the probe is placed should be at least 0.5 m in diameter so as to represent an "infinite" medium. Calibration to determine C background can be done by a series of ten 1 min readings, to be averaged, or by a single 1 h reading. C soil is determined from averaging several soil readings at a particular depth/location. For calibration purposes, it is best to take three samples around the access tube and to average the water contents corresponding to the average CR calculated for that depth. A minimum of five different water contents should be evaluated for each depth. Although some calibration curves may be similar, a separate calibration for each depth should be conducted. The lifetime of most probes is more than 10 years.

Gamma-ray attenuation: Whereas the neutron method measures the volumetric water content in a large sphere, gamma-ray absorption scans a thin layer. The dual-probe gamma device is nowadays mainly used in the laboratory since dielectric methods became operational for field use. Another reason for this is that gamma rays are more dangerous to work with than neutron scattering devices, as well as the fact that the operational costs for the gamma rays are relatively high.

Changes in gamma attenuation for a given mass absorption coefficient can be related to changes in total soil density. As the attenuation of gamma rays is due to mass, it is not possible to determine water content unless the attenuation of gamma rays due to the local dry soil density is known and remains unchanged with changing water content. Determining accurately the soil water content from the difference between the total and dry density attenuation values is therefore not simple. Compared to neutron scattering, gamma-ray attenuation has the advantage of allowing accurate measurements at a few centimetres below the air-surface interface. Although the method has a high degree of resolution, the small soil volume evaluated will exhibit more spatial variation due to soil heterogeneities.

7.2.3 Soil Water Dielectrics

When a medium is placed in the electric field of a capacitor or waveguide, its influence on the electric forces in that field is expressed as the ratio between the forces in the medium and the forces which would exist in vacuum. This ratio, called permittivity or "dielectric constant", is for liquid water about 20 times larger than that of average dry soil, because water molecules are permanent dipoles. The dielectric properties of ice, and of water bound to the soil matrix, are comparable to those of dry soil. Therefore, the volumetric content of free soil water can be determined from the dielectric characteristics of wet soil by reliable, fast, non-destructive measurement methods, without the potential hazards associated with radioactive devices. Moreover, such dielectric methods can be fully automated for data acquisition. At present, two methods which evaluate soil water dielectrics are commercially available and used extensively, namely time-domain reflectometry and frequency-domain measurement.

Time-domain reflectometry

Time-domain reflectometry is a method which determines the dielectric constant of the soil by monitoring the travel of an electromagnetic pulse, which is launched along a waveguide formed by a pair of parallel rods embedded in the soil. The pulse is reflected at the end of the waveguide and its propagation velocity, which is inversely proportional to the square root of the dielectric constant, can be measured well by actual electronics.

The most widely used relation between soil dielectrics and soil water content was experimentally summarized by Topp et al. (1980) as follows:

$$\theta_v = -0.053 + 0.029\varepsilon - 5.5 \cdot 10^{-4}\varepsilon^2 + 4.3 \cdot 10^{-6}\varepsilon^3 \tag{8}$$

where ε is the dielectric constant of the soil water system. This empirical relationship has proved to be applicable in many soils, roughly independent of texture and gravel content. However, soil-specific calibration is desirable for soils with low density or with a high organic content. For complex soil mixtures, the De Loor equation has proved useful.

Generally, the parallel probes are separated by 5 cm and vary in length from 10 to 50 cm; the rods of the probe can be of any metallic substance. The sampling volume is essentially a cylinder of a few centimetres in radius around the parallel probes. The coaxial cable from the probe to the signal-processing unit should not be longer than about 30 m. Soil water profiles can be obtained from a buried set of probes, each placed horizontally at a different depth, linked to a field data logger by a multiplexer.

Frequency-domain measurement

While time-domain reflectometry uses microwave frequencies in the gigahertz range, frequency-domain sensors measure the dielectric constant at a single microwave megahertz frequency. The microwave dielectric probe utilizes an open-ended coaxial cable and a single reflectometer at the probe tip to measure amplitude and phase at a particular frequency. Soil measurements are referenced to air, and are typically calibrated with dielectric blocks and/or liquids of known dielectric properties. One advantage of using liquids for calibration is that a perfect electrical contact between the probe tip and the material can be maintained. As a single, small probe tip is used, only a small volume of soil is ever evaluated, and soil contact is therefore critical. As a result, this method is excellent for laboratory or point measurements, but is likely to be subject to spatial variability problems if used on a field scale.

7.3 SOIL WATER POTENTIAL INSTRUMENTATION

The basic instruments capable of measuring matric potential are sufficiently inexpensive and reliable to be used in field-scale monitoring programmes. However, each instrument has a limited accessible water potential range. For example, tensiometers work well only in wet soil, while resistance blocks do better in moderately dry soil.

7.3.1 Tensiometers

The most widely used and least expensive water potential measuring device is the tensiometer. Tensiometers are simple instruments, usually consisting of a porous ceramic cup and a sealed plastic cylindrical tube connecting the porous cup to some pressure-recording device at the top of the cylinder. They measure the matric potential, because solutes can move freely through the porous cup. The tensiometer establishes a quasi-equilibrium condition with the soil water system. The porous ceramic cup acts as a membrane through which water flows, and therefore must remain saturated if it is to function properly. Consequently, all the pores in the ceramic cup and the cylindrical tube are initially filled with de-aerated water. Once in place, the tensiometer will be subject to negative soil water potentials, causing water to move from the tensiometer into the surrounding soil matrix. The water movement from the tensiometer will create a negative potential or suction in the tensiometer cylinder which will register on the recording device. For recording, a simple U-tube filled with water and/or mercury, a Bourdon-type vacuum gauge or a pressure transducer is suitable.

If the soil water potential increases, water moves from the soil back into the tensiometer, resulting in a less negative water potential reading. This exchange of water between the soil and the tensiometer, as well as the tensiometer's exposure to negative potentials, will cause dissolved gases to be released by the solution, forming air bubbles. The formation of air bubbles will alter the pressure readings in the tensiometer cylinder and will result in faulty readings. Another limitation is that the tensiometer has a practical working limit of $\psi \approx -85$ kPa. Beyond -100 kPa (≈ 1 atm), water will boil at ambient temperature, forming water vapour bubbles which destroy the vacuum inside the tensiometer cylinder. Consequently, the cylinders occasionally need to be de-aired with a handheld vacuum pump and then refilled.

Under drought conditions, appreciable amounts of water can move from the tensiometer to the soil. Thus, tensiometers can alter the very condition they were designed to measure. Additional proof of this process is that excavated tensiometers often have accumulated large numbers of roots in the proximity of the ceramic cups. Typically, when the tensiometer acts as an "irrigator", so much water is lost through the ceramic cups that a vacuum in the cylinder cannot be maintained, and the tensiometer gauge will be inoperative. Before installation, but after the tensiometer has been filled with water and degassed, the ceramic cup must remain wet. Wrapping the ceramic cup in wet rags or inserting it into a container of water will keep the cup wet during transport from the laboratory to the field. In the field, a hole of the appropriate size and depth is prepared. The hole should be large enough to create a snug fit on all sides, and long enough so that the tensiometer extends sufficiently above the soil surface for de-airing and refilling access. Since the ceramic cup must remain in contact with the soil, it may be beneficial in stony soil to prepare a thin slurry of mud from the excavated site and to pour it into the hole before inserting the tensiometer. Care should also be taken to ensure that the hole is backfilled properly, thus eliminating any depressions that may lead to ponded conditions adjacent to the tensiometer. The latter precaution will minimize any water movement down the cylinder walls, which would produce unrepresentative soil water conditions. Only a small portion of the tensiometer is exposed to ambient conditions, but its interception of solar radiation may induce thermal expansion of the upper tensiometer cylinder. Similarly, temperature gradients from the soil surface to the ceramic cup may result in thermal expansion or contraction of the lower cylinder. To minimize the risk of temperature-induced false water potential readings, the tensiometer cylinder should be shaded and constructed of non-conducting materials, and readings should be taken at the same time every day, preferably in the early morning. A new development is the osmotic tensiometer, where the tube of the meter is filled with a polymer solution in order to function better in dry soil.

7.3.2 Resistance Blocks

Electrical resistance blocks, although insensitive to water potentials in the wet range, are excellent companions to the tensiometer. They consist of electrodes encased in some type of porous material that within about two days will reach a quasi-equilibrium state with the soil. The most common block materials are nylon fabric, fiberglass and gypsum, with a working range of about –50 kPa (for nylon) or –100 kPa (for gypsum) up to –1 500 kPa. Typical block sizes are 4 cm × 4 cm × 1 cm. Gypsum blocks last a few years, but less in very wet or saline soil.

This method determines water potential as a function of electrical resistance, measured with an alternating current bridge (usually ≈ 1 000 Hz) because direct current gives polarization effects. However, resistance decreases if soil is saline, falsely indicating a wetted soil. Gypsum blocks are less sensitive to soil saltiness effects because the electrodes are consistently exposed to a saturated solution of calcium sulphate. The output of gypsum blocks must be corrected for temperature.

Because resistance blocks do not protrude above the ground, they are excellent for semi-permanent agricultural networks of water potential profiles, if installation is careful and systematic (WMO, 2001). When installing the resistance blocks it is best to dig a small trench for the lead wires before preparing the hole for the blocks, in order to minimize water movement along the wires to the blocks. A possible field problem is that shrinking and swelling soil may break contact with the blocks. On the other hand, resistance blocks do not affect the distribution of plant roots.

Resistance blocks are relatively inexpensive. However, they need to be calibrated individually. This is generally accomplished by saturating the blocks in distilled water and then subjecting them to a predetermined pressure in a pressure-plate apparatus, at least at five different pressures before field installation. Unfortunately, the resistance is less on a drying curve than on a wetting curve, thus generating hysteresis errors in the field because resistance blocks are slow to equilibrate with varying soil wetness. As resistance-block calibration curves change with time, they need to be calibrated before installation and to be checked regularly afterwards, either in the laboratory or in the field.

7.3.3 Psychrometers

Psychrometers are used in laboratory research on soil samples as a standard for other techniques, but a field version is also available, called the Spanner psychrometer. This consists of a miniature thermocouple placed within a small chamber with a porous wall. The thermocouple is cooled by the Peltier effect, condensing water on a wire junction. As water evaporates from the junction, its temperature decreases and a current is produced which is measured by a meter. Such measurements are quick to respond to changes in soil water potential, but are very sensitive to temperature and salinity.

The lowest water potential typically associated with active plant water uptake corresponds to a relative humidity of between 98% and 100%. This implies that, if the water potential in the soil is to be measured accurately to within 10 kPa, the temperature would have to be controlled to better than 0.001 K. This means that the use of field psychrometers is most appropriate for low matric potentials, of less than –300 kPa. In addition, the instrument components differ in heat capacities, so diurnal soil temperature fluctuations can induce temperature gradients in the psychrometer. Therefore, Spanner psychrometers should not be used at depths of less than 0.3 m, and readings should be taken at the same time each day, preferably in the early morning. In summary, soil psychrometry is a difficult and demanding method, even for specialists.

7.4 SITE SELECTION AND SAMPLE SIZE

There is no standard depth or measurement interval at which soil moisture observations are taken, since this strongly depends on the research objectives for which the sensors are installed. The International Soil Moisture Network provides an extensive database with harmonized in situ soil moisture time series of networks all over the world. Here the data are harmonized to a half-hourly measurement interval whenever possible. Most networks and stations in the ISMN measure soil moisture at several depths, from 0.05 m up to 0.50

m or 1 m. As a result, the behaviour of soil moisture at different depths can be compared and used to validate measurements. Measurements of other meteorological parameters are very valuable for determining soil moisture. For example, precipitation data at the measurement site can help validate the soil moisture data.

The representativeness of any soil moisture observation point is limited because there are likely to be significant variations, both horizontally and vertically, in the soil structure (porosity, density, and chemical composition), land cover and relief. It is pivotal that soil moisture and its variability be captured on the scale necessary for conducting studies on hydrological processes and for satellite validation. Gravimetric water content determinations or indirect measurements of soil moisture are only reliable at the point of measurement, making it necessary to collect a large number of samples to adequately describe the site's soil moisture status. In order to estimate the number of samples n needed at a local site to determine soil water content at an observed level of accuracy (L), the following equation can be used:

$$n = 4\left(\sigma^2 / L^2\right) \quad (9)$$

where σ_2 is the sample variance generated from a preliminary sampling experiment. For example, suppose that a preliminary sample yielded a (typical) σ_2 of 25% and the accuracy level needed to be within 3%, 12 samples would be required from the site (if it can be assumed that water content is normally distributed across the site). Upscaling the point measurements obtained by gravimetric water content determination or indirect measurements with in situ sensors has been the subject of many studies. Upscaling methods vary from relatively straightforward interpolation and time/rank stability techniques to more complicated techniques such as statistical transformations and land surface modelling.

7.5 REMOTE-SENSING OF SOIL MOISTURE

A single observation location cannot provide absolute knowledge of regional soil moisture. Soil moisture is highly variable in both space and time, rendering it difficult to measure on the continental or global scale needed by researchers. Space-based remote-sensing of soil moisture accommodates these needs by providing surface soil moisture observations on a global scale every one to two days under a variety of conditions.

In general, remote-sensing aims to measure properties of the Earth's surface by analysing the interactions between the ground and EMR. This can be done by recording the naturally emitted radiation (passive systems) or by illuminating the ground and recording the reflecting signal (active systems). Soil moisture is usually assessed through its effects on the soil's electric or thermal properties. While microwave remote-sensing observations are sensitive to the soil's dielectric constant, IR remote-sensing systems are sensitive to its thermal conditions.

Over the last decades, many soil moisture datasets have been developed from various space-borne instruments using different retrieval algorithms. Recently, several of these datasets from both active and passive microwave remote-sensing observations have been combined, generating a global soil moisture dataset covering the last 30 years. Although remote-sensing has proven to be a valuable tool to measure soil moisture on a global scale, in situ measurements are imperative for the calibration and validation of satellite-based soil moisture retrievals. ISMN, a global in situ soil moisture database, was mainly developed for the validation of satellite products. Many validation efforts have been undertaken to assess the quality of remote-sensing products using *in situ* measurements. In addition, many studies have focused on error characterization of the different soil moisture products. These studies show that most soil moisture products from remote-sensing are capable of depicting seasonal and short-term soil moisture changes quite well. However, biases in the absolute value and dynamic range may be large when compared to in situ and modelled soil moisture data.

7.5.1 Microwave Remote-Sensing

Microwave remote-sensing uses electromagnetic waves with wavelengths of 1 m to 1 cm, which corresponds to frequencies of 0.3 to 300 GHz. An important quality of these microwaves is that they can travel through the Earth's atmosphere undisturbed and thus allow observations to be made independent from cloud coverage. Furthermore, since they are not bound to illumination by the sun, microwave measurements are operable all day long.

When applied to remote-sensing of the Earth's surface, Kirchhoff's radiation law states that the emission of a body is equal to one minus its reflectivity. This means that emission and reflection are complementary, and thus that surfaces that are good scatters are weak emitters and vice versa. As a result, active and passive microwave systems are influenced inversely by the same physical phenomena on the ground. Fresnel's reflection law describes the relationship between the dielectric constant and reflectivity (and thus emissivity), where a higher dielectric constant yields a higher rate of reflection (and smaller emissivity). At microwave lengths, the dielectric constant of water is of an order of magnitude larger than that of dry soils. Therefore, the dielectric constant of soils rises with increasing soil moisture. With these physical relations, it is possible to retrieve soil moisture of the Earth's surface from passive as well as from active microwave remote-sensing systems.

Microwave beams are able to interact to some extent with the volumes of targets since their waves are longer and are not reflected immediately at the surface. Thus, information about the inner conditions of vegetation or soils, for example, can be gained. As a rule of thumb, the longer the wavelength, the deeper the radiation penetrates into volumes. In contrast, optical waves only interact with surfaces and tell us about the visible colour and brightness.

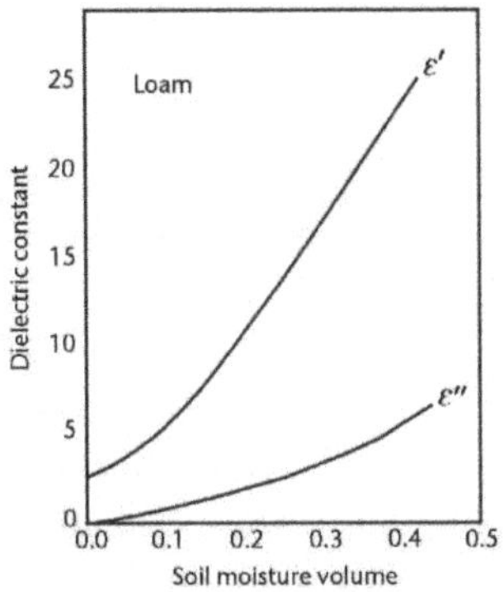

Fig. 1: The relationship between the complex dielectric constant (ε' and ε" are magnitudes of the real and imaginary parts, respectively) and volumetric soil moisture for a loamy soil at a frequency of 5 GHz .

When observed from above the canopy, vegetation affects the microwave emission in two ways: first, vegetation absorbs or scatters the radiation emitted from the soil; second, the vegetation also emits its own radiation. Under a sufficiently dense canopy, the emitted soil radiation will become totally masked and the observed radiation will be mostly due to vegetation. Generally, all frequency bands used in microwave remote-sensing of soil moisture are sensitive to vegetation and require some correction in the data for this. Higher-frequency bands are more vulnerable to vegetation influences.

7.5.2 Multi-Frequency Radiometers

Passive systems like radiometers record the brightness temperature of the Earth's surface. Brightness temperatures are related to the amount of emissivity (and thus reflection) described by the Rayleigh-Jeans approximation of Planck's law. This law states that brightness temperatures are a function of the physical temperature and emissivity. The amount of emission depends on the dielectric constant of the emitting body as described by Fresnel's reflection law.

Since 1978, instruments have been providing global passive data over land and oceans (Figure 2), beginning with the Scanning Multichannel Microwave Radiometer (1978–1987), the Special Sensor Microwave – Imager (since 1987), the Tropical Rainfall Measuring Mission (since 1997) and, more recently, the Advanced Microwave Scanning Radiometers (AMSR-E, 2002–2011 and AMSR-2 since 2012), the Coriolis WindSat (since 2003) and the Chinese satellites FengYun-3 (since 2010). Initially these instruments were not designed for soil moisture observations but for precipitation, evaporation, sea-surface temperatures and cryospheric parameters. However, in the 1970s studies already showed the potential of retrieving soil advantage of radiometers is that data are available from multiple multi-frequency microwave radiometers since 1978, providing a long-term dataset to investigate trends and anomalies.

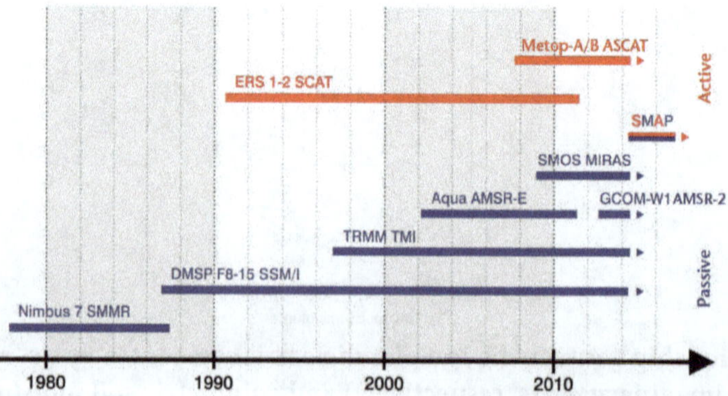

Fig. 2: Active and passive microwave sensors used for soil moisture retrieval

The instruments used for soil moisture remote-sensing have frequencies varying from 6.6 to 10.7 GHz. It has to be taken into account that a higher microwave frequency leads to less accurate estimates of soil moisture since attenuation due to vegetation increases and penetration ability decreases. Therefore, retrievals from the Scanning Multichannel Microwave Radiometer (6.6 GHz), AMSR-E (6.9 GHz), WindSat (6.8 GHz) and AMSR-2 (6.9 GHz) tend to be more accurate. Another advantage of these sensors is that spatial resolution and radiometric accuracy are much improved. The spatial resolution of AMSR-E is 56 km where the soil moisture products are provided at a spatial resolution of 0.25°.

7.5.3 Scatterometers

A scatterometer is an active microwave instrument (AMI) that continuously transmits short directional pulses of energy towards the Earth's surface and detects the returned energy. The amount of energy returned to the instrument depends upon geometric and dielectric properties of the surface and is often referred to as normalized radar cross-section or backscatter (sigma nought, σ_0). Sacrificing range and spatial resolution, scatterometers surpass other types of radars in accuracy and stability for measuring the radar cross-section of a target. Space-borne scatterometers were initially developed and designed to derive wind speed and direction over the oceans. Nevertheless, a number of studies acknowledge the capacity of scatterometers for land applications such as soil moisture monitoring. Since European scatterometers operate in longer wavelengths (5.3 GHz) than those of the United States (14 GHz), they are more suitable for soil moisture retrieval.

The unique instrument design of the European scatterometers on board the European Remote-sensing (ERS) satellites and the Meteorological Operational (MetOp) satellites enables soil moisture retrieval on a global scale with almost daily coverage. Both scatterometers, the AMI in wind mode on board ERS and the Advanced Scatterometer (ASCAT) on board MetOp, operate in C-band (5.3 GHz) with a wavelength of approximately 5.6 cm. The major differences between these two scatterometers are the number of sideways-looking antennas and the range of the incidence angles observed. The spatial resolution of the AMI is approximately 50 km while the ASCAT product is provided at spatial resolutions of 25 km and 50 km.

7.5.4 Synthetic Aperture Radars

Space or airborne synthetic aperture radars (SAR) are active microwave sensor systems that offer a higher spatial resolution than scatterometers due to advanced signal processing. As side-looking imaging radars, they operate similarly to scatterometers and use the same frequency domain. Besides hydrological applications, SAR systems can be used for the accurate retrieval of three-dimensional geometries, as they enable interferometry.

As a side-looking imaging radar moves along its path on the ground, it accumulates data. The spatial resolution of radars is dependent on the (limited) physical size of its antenna, the aperture. Taking advantage of the along-track motion of the carrier, an SAR system simulates a bigger synthetic aperture as it records amplitude and phase of the ground targets continuously while they are visible to the SAR. These multiple measurements of each target are then summed up coherently. Smaller objects are subsequently resolved on the ground. However, higher energy consumption and a smaller footprint result in a longer revisit time on individual locations and thus the temporal resolution of SARs is inferior to other microwave systems these days.

The higher complexity of soil and surface properties at the scale below 10 km introduces additional error and uncertainty sources. As a consequence, SAR systems are not yet employed in operational soil moisture services but instead are used for pre-operational services and scientific products. Nonetheless, upcoming SAR satellite missions such as the European Space Agency (ESA) Sentinel1 programme promise improved temporal and radiometric resolution and a suggestion has been made to use SARs for operational soil moisture services on a local scale.

7.5.5 Dedicated L-Band Missions

As stated before, lower frequencies tend to be less sensitive to vegetation interactions and are therefore thought to be more suitable for soil moisture retrieval. Hence, the first two spaceborne missions specifically designed for soil moisture retrieval operate in the L-band channel (1.4 GHz). The aim of the Soil Moisture and Ocean Salinity (SMOS) and the Soil Moisture Active Passive (SMAP) missions is to provide absolute soil moisture with a maximum root mean square (RMS) error of 0.04 m_3/m_3.

The SMOS mission of ESA was launched successfully on 2 November 2009. The instrument on board the SMOS satellite has a unique design to provide the spatial resolution needed for measuring soil moisture. The so-called Microwave Imaging Radiometer using Aperture Synthesis (MIRAS) is a 2D interferometric radiometer, on which the size of the antenna needed for measuring at the required spatial resolution is simulated through 69 small antennas. MIRAS provides brightness temperatures with a spatial resolution varying between 30 and 50 km. Global coverage is achieved every 2–3 days.

The SMAP mission, run by the United States National Aeronautics and Space Administration (NASA), was launched on 31 January 2015. Like SMOS, the passive microwave instrument operates in Lband to enhance the sensitivity to soil moisture. However, the instrument design for SMAP is very different from SMOS. SMAP uses a real aperture antenna in the shape of a large (6 m) parabolic reflector that rotates. Measurements are made with a spatial resolution of 40 km. In addition to the passive measurements, SMAP also carries a radar that makes concurrent measurements at a spatial resolution of 1–3 km. By combining active and passive measurements, SMAP provides a soil moisture product with a spatial resolution of 10 km.

7.5.6 Soil Moisture Retrieval

For retrieving soil moisture, it is necessary to have models that are capable of accounting for vegetation and surface roughness effects on the microwave signal and then to convert accordingly the received intensity to soil moisture values. Again, it should be noted that a shorter wavelength leads to inferior performance due to vegetation scattering and less penetration depth. Soil moisture retrieval is not possible over densely vegetated areas such as vegetation canopy. Additionally, retrieved estimates of soil moisture are only reasonable over snow-free and non-frozen soils.

Passive systems measure the microwave brightness temperature and derive indirectly the emissivity, which is then ingested into a radiative transfer model. Data on soil temperature, roughness, texture and other parameters of the observed area are necessary ancillary information. Data from passive microwave observations are available from AMSR-E using either the Vrije Universiteit Amsterdam (VUA)–NASA retrieval algorithm developed by VUA and NASA and based on the land parameter retrieval model, the official NASA AMSR-E product or the retrieval algorithm from the University of Montana. All of these retrieval algorithms are based on radiative transfer equations. However, the retrieval algorithms vary significantly and generate quite different soil moisture values. The VUA–NASA retrieval algorithm solves for vegetation optical depth and the soil dielectric constant simultaneously. Soil moisture is calculated using the Wang–Schmugge mixing model.

The SMOS instrument provides an operational soil moisture product. The SMOS retrieval algorithm uses an iterative approach to minimize the cost function between modelled brightness temperatures and the direct measurements. In this way the best set of parameters is found, including the soil moisture and vegetation.

Active instruments measure the backscattered intensity, which is a function of roughness, incidence angle and dielectric properties of the surface. Again, vegetation and other influences contribute to the signal, which is used to determine the backscatter coefficient. Soil moisture retrieval provided as an operational product from the ASCAT instrument and as a scientific product from the AMI in wind mode relies on a semi-empirical change-detection method. This method, the TU Wien change detection algorithm, is tailored to the unique instrument design. Assuming a linear relationship between radar backscatter and soil moisture, in the decibel domain, a relative measure of moisture in the first few centimetres of soil can be obtained, representing the degree of saturation (0%–100%). In very dry regions, particularly over sand deserts, the retrieval approach fails, seemingly due to a complex mechanism of surface, volume and sub-surface scattering.

An overview of operational soil moisture products is given in the table below.

Operational soil moisture products and their characteristics

Product reference	SMOS	AMSR-E	ASCAT
Satellite			
Name	SMOS	Aqua	MetOp-A/B
Agencies	ESA/CNES[a]/CDTI[b]	NASA	EUMETSAT/ESA
Lifetime	Since 2.11.09	4.5.02 – 4.10.11	Since 19.10.06
Orbit	Polar	Polar	Polar
Altitude	758 km	705 km	837 km
Period	100 min	99 min	100 min
Equator crossing time	6 a.m. (ascending)6 p.m. (descending)	1.30 p.m. (ascending)1.30 a.m. (descending)	9.30 p.m. (ascending)9.30 a.m. (descending)
Type	Research satellite	Research satellite	Operational (3 satellites)
Product reference	SMOS	AMSR-E	ASCAT
Sensor			
Name	MIRAS	AMSR-E	ASCAT
Type	Synthetic aperture radiometer	Multi-frequency real aperture radiometer	Real aperture scatterometer
Swath	1 000 km	1 450 km	2 x 550 km
Scanning principle	Forward looking2D interferometer	Rotating parabolic reflector	6 side-looking fan beam antennas
Incidence angle range	0°– 55°	55°	25°– 53° (mid beam); 34°– 64° (fore and aft beams)
Frequency	1.4 GHz	6.9, 10.7, 18.7, 23.8, 36.5, and 89 GHz	5.3 GHz
Polarization	H and V (polarimetric mode optional)	H and V	VV

Product reference	SMOS	AMSR-E	ASCAT
Spatial resolution	30–50 km	75 x 43 km at 6.9 GHz	25/50 km
Daily global coverage	~82%	~90%	~82%
Retrieval			
Model name	L-MEB	Land parameter retrieval model	WARP
Forward model	Radiative transfer model	Radiative transfer model	Semi-empirical change-detection
Model complexity	High	Medium	Low
Inversion approach	Iterative least-square matching	Iterative least-square matching	Direct inversion
Concurrent retrievals	Soil temperature, vegetation optical depth, roughness	Soil temperature, vegetation optical depth	None
Model calibration	None	None	Based on long-term time series
Need for auxiliary data	High	Medium	Low
Error propagation estimates	Not available	Available	Available
Product			
Target quantity	Volumetric soil moisture	Volumetric soil moisture	Degree of saturation
Units	$m^3 m^{-3}$	$m^3 m^{-3}$	0–1 or %
Grid	Fixed ISEA4-9 Discrete Global Grid	Regular grid	Swath geometry
Pixel spacing	15 km	0.25°	12.5 km
Data latency	Within a few days after sensing	Irregular updates	Within 130 min after sensing
Notes: National Centre for Space Studies (France) Centro para el Desarrollo Tecnológico Industrial (Spain)			

7.5.7 Thermal Infrared Remote-Sensing

All bodies with a temperature above absolute zero emit electromagnetic energy in the thermal IR domain. By detecting the thermal properties of the Earth's surface, soil moisture can be derived based on the distinct differences in thermal properties of soil and water. Thermal IR remote-sensing has been used in an increasing number of studies for the derivation of soil moisture. The advantage of thermal IR remote-sensing is that it can provide soil moisture information on a spatial resolution down to a few metres. Furthermore, it can provide soil moisture information over dense vegetation, which is one of the limitations of microwave remote-sensing. The disadvantages of thermal IR remote-sensing are that it is unable to measure soil moisture when cloud cover is present and it is considerably affected by atmospheric phenomena. Therefore, complex noise-removal mechanisms are needed in most cases. Thermal IR remote-sensing of soil moisture is not as straightforward as microwave remote-sensing since there is no direct link between temperature data and soil moisture. Nevertheless, several approaches exist to indirectly retrieve soil moisture data using thermal IR observations from the geostationary operational environmental satellite (GOES), advanced very high resolution radiometer (AVHRR), moderate resolution imaging spectroradiometer, Landsat and others.

The first approach is called the triangle approach and is based on the empirical relationship between soil moisture, soil temperature and fractional vegetation cover. This relationship was demonstrated and resulted in a triangular scatterplot of surface temperatures and the remotely sensed normalized difference vegetation index.

The second approach makes use of the differences in thermal properties between water and soils. Water differs from many other matters in its relatively large heat capacity and thermal inertia. Thermal inertia is defined as the resistance of an object against its heating for 1 K. The thermal inertia of water is relatively high, which indicates a high resistance to temperature changes. It has been shown that the behaviour of land surface temperature in the morning strongly depends on soil moisture in the soil, since the water will heat up more slowly. One of the approaches that uses this behaviour is the calculation of the apparent thermal inertia (ATI), which can be done when measuring the difference between maximum and minimum temperatures over one day. It is described as:

$$\mathrm{ATI} = (1 - A) / \Delta T \tag{10}$$

Where A is the albedo of the pixel in the visible band and ΔT the difference between the minimum and maximum temperature. Many studies have already assessed the potential of ATI to describe soil moisture and its spatial and temporal variability.

Another method to retrieve soil moisture using thermal IR remote-sensing is by integrating the data into land surface models. Soil moisture controls latent heat fluxes by way of both evaporation and transpiration, where wet soil conditions lead to increased evaporation and transpiration. The atmosphere–land exchange inversion model (ALEXI) uses the relationship between evaporation, transpiration and soil moisture to derive soil moisture data. All major components, including the latent heat flux, of the energy budget are estimated from net radiation and vegetation parameters retrieved from AVHRR and GOES. Accordingly, soil moisture can be derived from latent heat fluxes by using a soil water stress function. An intercomparison of soil moisture retrieved from microwave remote-sensing and ALEXI showed that the two datasets are complementary: ALEXI is better at estimating soil moisture over dense vegetation and microwave remote-sensing shows more reliable results over low to moderate vegetation.

8

AGRO-CLIMATIC ZONES AND CLIMATIC NORMAL

8.1 INTRODUCTION

Food and Agriculture Organization (FAO) defined an agro-climatic zone (ACZ) as a land unit represented accurately or precisely in terms of major climate and growing period, which is climatically suitable for certain range of crops and cultivars. In other words, it is an extension of the climate classification keeping in view the suitability to agriculture.

8.2 AGRO-CLIMATIC ZONES

- The Planning Commission of India has divided the India into fifteen broad agro-climatic zones on the basis of physiography, soils, geological formation, climate, cropping patterns, and development of irrigation and mineral resources for extensive agricultural planning and developing future strategies.
- The main objective was to integrate plans of the agro-climatic regions with the state and national plans to enable policy development based on techno-agro-climatic considerations.
- The agro-climatic zone division helps to achieve the optimum utilization in a suitable manner within the framework of resource constraints and potentials of each region. This classification also helps in the development of efficient farming systems and crop improvement research programmes, and establishing principles for improved resource management.
- India was divided into 15 Agroclimatic zones/region.

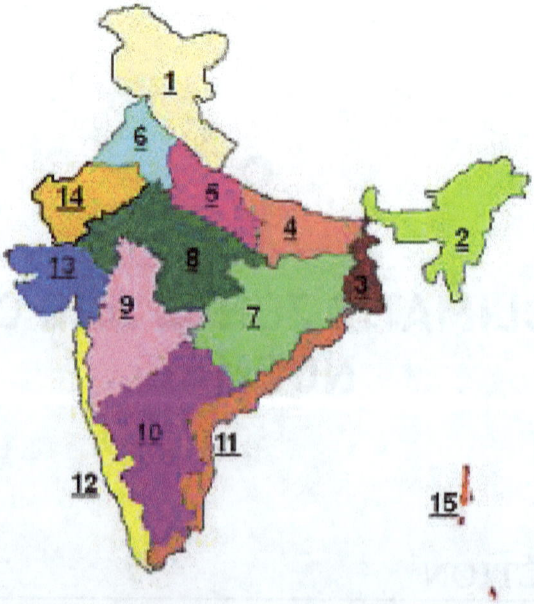

Figure 13.1 : Agroclimatic zones of India

1. **Western Himalayan Region**
   b. **Area:** Jammu and Kashmir, Himachal Pradesh and the hill region of Uttarakhand
   c. **Major Crops:** Saffron, maize, barley, oats and wheat; peaches, apricot, pears, cherry, almond, litchis, walnut
2. **Eastern Himalayan Region**
   a. **Area:** Arunachal Pradesh, the hills of Assam, Sikkim, Meghalaya, Nagaland, Manipur, Mizoram, Tripura, and the Darjeeling district of West Bengal
   b. **Major Crops:** Rice, maize, potato, and tea; orchards of pineapple, litchi, oranges and lime.
3. **Lower Gangetic Plain Region**
   a. **Area:** West Bengal (except the hilly areas), eastern Bihar and the Brahmaputra valley
   b. **Major Crops:** Rice, jute, maize, potato, and pulses
4. **Middle Gangetic Plain Region**
   a. **Area:** Parts of Uttar Pradesh and Bihar
   b. **Major Crops:** Rice, maize, millets, wheat, gram, barley, peas, mustard and potato
5. **Upper Gangetic Plains Region**
   a. **Area:** Central and western parts of Uttar Pradesh and the Hardwar and Udham Nagar districts of Uttarakhand
   b. **Major Crops:** Wheat, rice, sugarcane, millets, maize, gram, barley, oilseeds, pulses and cotton.

6. **Trans-Ganga Plains Region**
   a. **Area:** Punjab, Haryana, Chandigarh, Delhi and the Ganganagar district of Rajasthan.
   b. **Major Crops:** wheat, sugarcane, cotton, rice, gram, maize, millets, pulses and oilseeds.
7. **Eastern Plateau and Hills**
   a. **Area:** Chhotanagpur Plateau, extending over Jharkhand, Odisha, Chhattisgarh and Dandakaranya.
   b. **Major Crops:** Rice, millets, maize, oilseeds, ragi, gram, potato, tur, groundnut, soyabean, urad, castor, and groundnut.
8. **Central Plateau and Hills**
   a. **Area:** Bundelkhand, Baghelkhand, Bhander Plateau, Malwa Plateau, and Vindhyachal Hills
   b. **Major Crops:** millets, wheat, gram, oilseeds, cotton and sunflower.
9. **Western Plateau and Hills**
   a. **Area:** Southern part of Malwa plateau and Deccan plateau (Maharashtra).
   b. **Major Crops:** Wheat, gram, millets, cotton, pulses, groundnut, oilseeds, sugarcane, rice, wheat, oranges, grapes and bananas.
10. **Southern Plateau and Hills**
    a. **Area:** Interior Deccan and includes parts of southern Maharashtra, the greater parts of Karnataka, Andhra Pradesh, and Tamil Nadu uplands from Adilabad District in the north to Madurai District in the south.
    b. **Major Crops:** millets, oilseeds, pulses, coffee, tea, cardamom and spices.
11. **Eastern Coastal Plains and Hills**
    a. **Area:** Coromandal and northern Circar coasts of Andhra Pradesh and Orissa
    b. **Major Crops:** Rice, jute, tobacco, sugarcane, maize, millets, groundnut and oilseeds
12. **Western Coastal Plains and Ghats**
    a. **Area:** Malabar and Konkan coastal plains and the Sahyadris
    b. **Major Crops:** Rice, coconut, oilseeds, sugarcane, millets, pulses and cotton
14. **Gujarat Plains and Hills**
    a. **Area:** Hills and plains of Kathiawar, and the fertile valleys of Mahi and Sabarmati rivers
    b. **Major Crops:** Groundnut, cotton, rice, millets, oilseeds, wheat and tobacco
15. **Western Dry Region**
    a. **Area:** West of Aravalli (Rajasthan)
    b. **Major Crops:** Bajra, jowar, moth, wheat and gram
16. **Island Region**
    a. **Area:** Andaman-Nicobar and Lakshadweep
    b. **Major Crops:** rice, maize, millets, pulses, arecanut, turmeric and cassava.

Agro-Climatic Zones under NARP

- Under National Agricultural Research Project (NARP) of Indian Council of Agricultural Research (ICAR), the country was divided in 127 agro-climatic zones to address the location specific agricultural problems. ICAR reconstituted these agroclimatic zones and there are 131 agroclimatic zones.
- These agroclimatic zones were delineated based on topography, rainfall, temperature and cropping pattern. The administrative boundary of Taluka/Block was considered as a unit area.
- The state wise agroclimatic zones are given below.

Andhra Pradesh : There are 9 agroclimatic zones in the state, these are

1. North coastal zone
2. Godavari zone
3. Krishna zone
4. Southern zone
5. Scarce rainfall zone
6. High altitude and tribal zone

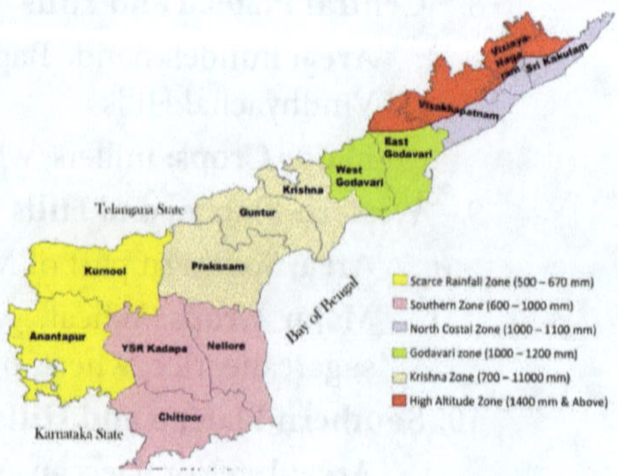

Assam : Six agroclimatic zones

1. North bank plain zone
2. Upper Brahmaputra valley zone
3. Central Brahmaputra valley zone
4. Lower Brahmaputra valley zone
5. Bank valley zone
6. Hill zone

Bihar : Three agroclimatic zones .

7. N-W alluvial zone
8. N-E alluvial zone
9. South Bihar alluvial plain zone

Chhattisgarh: Three agroclimatic zones

1. Northern hill zone
2. Chhattisgarh plain zone
3. Bastar plateau zone

Gujarat : Eight agroclimatic zones

1. South Gujarat (Heavy rainfall) zone
2. South Gujarat zone
3. Middle Gujarat zone
4. North Gujarat zone
5. North West Gujarat zone
6. North Saurashtra zone
7. South Saurashtra zone
8. Bhal and coastal zone

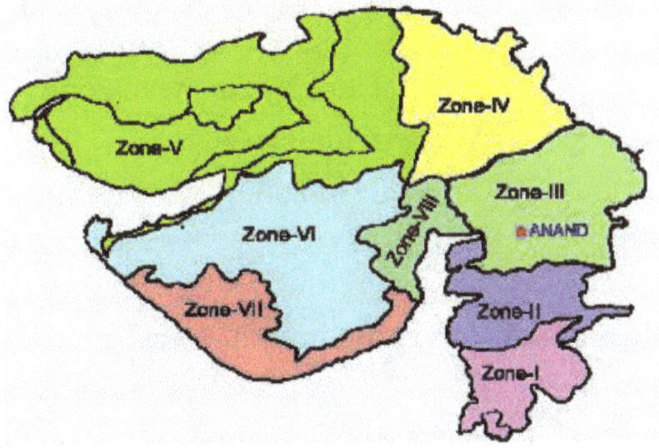

Haryana: Two agroclimatic zones

1. Western zone
2. Eastern zone

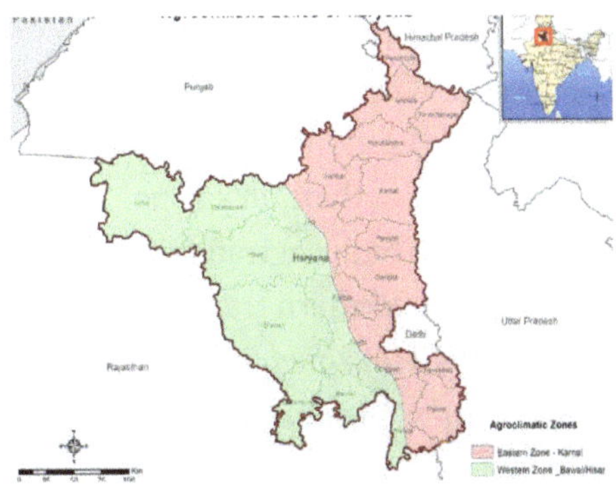

Himachal Pradesh: Four agro-climatic zones

1. Sub mountain and low hills
2. Mid-hill sub humid zone
3. High hills temperate wet zone
4. High hills temperate dry zone

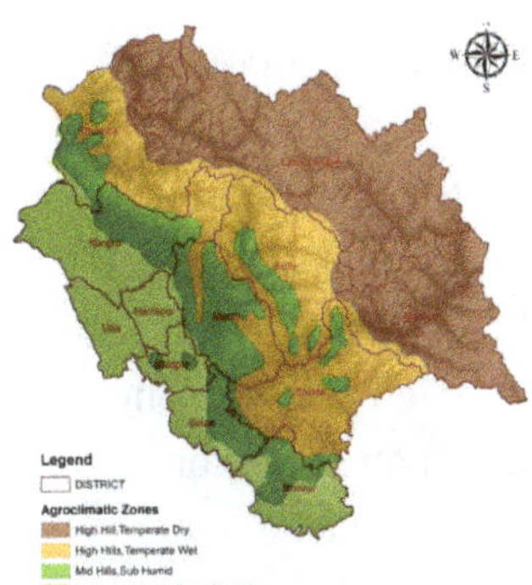

Jammu and Kashmir: Four agro-climatic zones

1. Low altitude subtropical zone (Jammu)
2. High altitude Intermediate zone
3. Temperate zone (Kashmir)
4. Cold arid zone (Ladhakh)

Jharkhand: Three agroclimatic zones

1. Central and N-E plateau zone
2. Western plateau zone
3. S-E plateau zone

Karnataka: Ten agroclimatic zones

1. N-E transition zone
2. N-E dry zone
3. Northern dry zone
4. Central dry zone
5. Eastern dry zone
6. Southern dry zone
7. Southern transition zone
8. Northern transition zone
9. Hill zone
10. Coastal zone

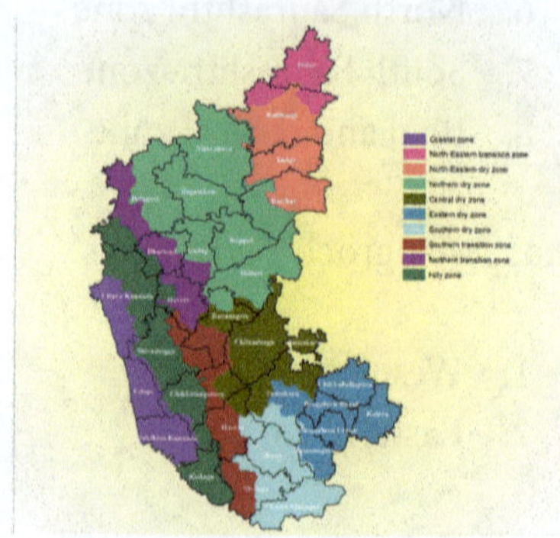

Kerala: Eight agroclimatic zones

1. Northern zone
2. Southern zone
3. Central zone
4. High altitude zone
5. Onattukura zone
6. Kuttanadkole zone
7. Pokkali zone problem are
8. Low rainfall zone

Madhya Pradesh: Eight agroclimatic zones

1. Kaimur Plateau and Satpuda hill zone
2. Vindhyan plateau
3. Narmada valley zone
4. High altitude zone
5. Onattukura zone
6. Kuttanadkole zone
7. Pokkali zone problem are
8. Low rainfall zone

Maharashtra: Nine agroclimatic zones

1. South Konkan Coastal zone
2. North Konkan Coastal zone
3. Western Ghat zone
4. Sub montane zone
5. Western Maharashtra plain zone
6. Scarcity zone
7. Central Maharashtra plateau zone
8. Central Vidarbha zone
9. Eastern Vidarbha zone

N-E Hill Zone: Six agroclimatic zones

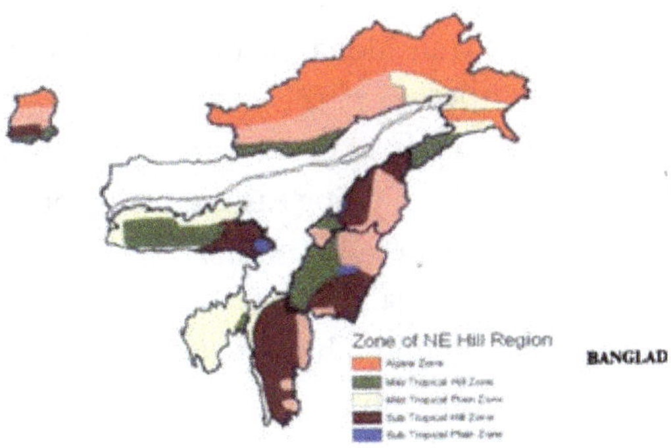

1. Alpine zone
2. Temperate sub-Alpine zone
3. Subtropical plain zone
4. Subtropical hill zone
5. Mild tropical plain zone
6. Mild tropical plain zone

Odisha: Ten agroclimatic zones

1. N-W plain zone
2. North Central plain zone
3. N-E plain zone
4. East and S-E coastal plain zone
5. N-E ghat zone
6. Eastern ghat high land zone
7. S-E ghat zone
8. Western undulating zone
9. West central table land zone
10. Mid- central table land zone

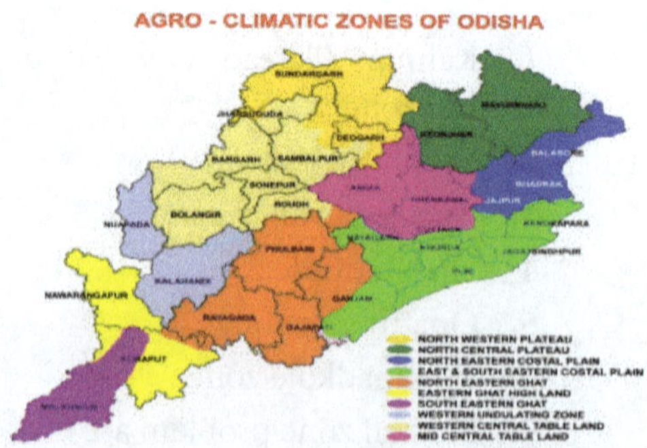

Punjab: Six agroclimatic zones

1. Sub mountain undulating region
2. Undulating plain region
3. Central plain region
4. Western plain region
5. Western region
6. Flood plain region

Rajasthan: Nine agroclimatic zones

1. Arid western plain zone
2. Irrigated N-W plain zone
3. Transitional plain zone of in-land drainage
4. Transitional plain zone of Luni basin
5. Semiarid eastern plain zone
6. Flood prone eastern plain zone
7. Subhumid southern plain and alluvial hill zone
8. Southern humid plain zone
9. S-E humid plain zone

Tamil Nadu: Seven agroclimatic zones

1. North Eastern zone
2. North Western zone
3. Western zone
4. Cauvery delta zone
5. Southern zone
6. High rainfall zone
7. Hill and high altitude zone

Telangana : Three agroclimatic zones

1. North Telangana Zone
2. Central Telangana Zone
3. South Telangana Zone

Uttar Pradesh: Nine agroclimatic zones

1. Bhabhar and Tarai Zone
2. Western Plains Zone
3. Mid Western Plains Zone
4. South Western Plains Zone
5. Central Plains Zone
6. Bundelkhand Zone
7. North Eastern Plains Zone
8. Eastern Plains Zone
9. Vindhya Zone

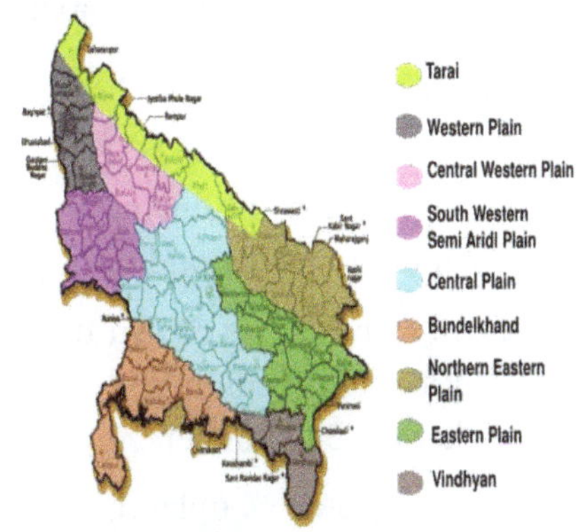

Uttarakhand: Two agroclimatic zones

1. Hill zone
2. Tarai zone

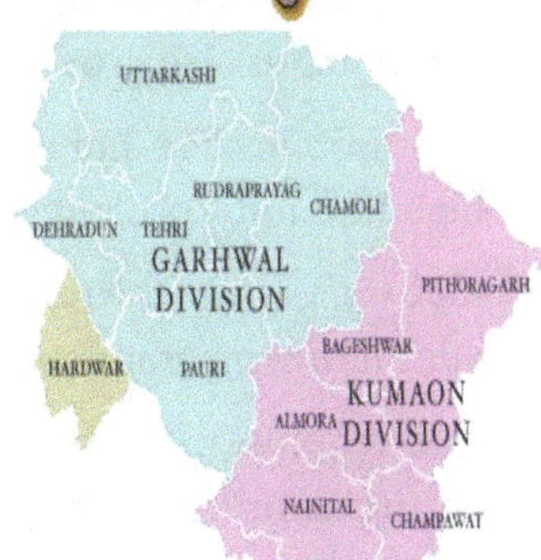

MAP OF UTTARAKHAND

West Bengal : Six agro-climatic zones

1. Hill Zone
2. Tarai zone
3. Old Alluvial zone
4. New Alluvial Zone
5. Red and Laterite soil zone
6. Coastal saline soil zone

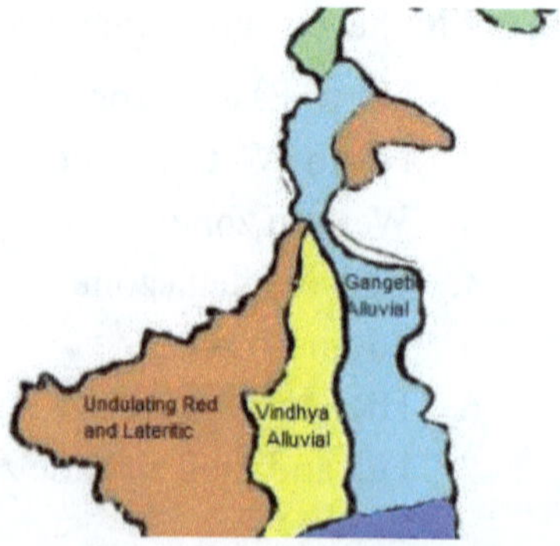

8.3 CLIMATIC NORMAL FOR CROPS

- Climate regulates and determines the suitability of crops for a region whereas; weather decides the growth, development and yield of crops grown in that region.
- Unless the crops and cultivars are well adapted to the areas where they are grown, their cultivation in that area is not economical.
- Weather induced variability in the crop productivity account for about 67% and the rest 33% are due to other factors including soil and nutrient management.
- Every crop requires a certain set of optimum climatic conditions relating to temperature, sunlight, moisture, wind etc. during different stages of their growth and development for maximum possible yield.

8.3.1 Rice

- Rice is a heat loving crop. It requires high temperature, ample water supply, and high humidity during the growth period. The capacity of rice to thrives well under flooded conditions gives rice a unique status
- In India rice is grown under widely varying conditions of altitude and climate. Rice cultivation in India extends from 8 to 35°N latitude and from sea level to as high as 3000 meters.
- It is grown as upland, lowland both under rainfed and irrigated conditions. In addition it is also grown in seasonally flooded wetland.
- Rice crop needs a hot and humid climate. It is best suited to regions which have high humidity, prolonged sunshine and an assured supply of water.
- Rice is grown in the regions having temperature range between 18 and 35^{0}C. The optimum temperature requirement of rice for its growth is 22-30^{0}C.
- Rice requires mean temperature not less than 15^{0}C, preferably more than 20^{0}C during entire growth period. Generally, 10^{0}C is taken as the lowest critical temperature to compute the heat unit requirement of rice crop.
- The temperature requirement varies with the growth stage of the crop. It should

be 20-22°C at the time of sowing, 23-25°C during growth and 25-30°C at the harvesting time.

- Low temperatures inhibits tillering and prolong crop duration. On an average, critical low temperature for inducing sterility is 15-17^{0}C. A temperature of 27-29^{0}C is ideal for flower initiation.
- Flowering is inhibited by low relative humidity. The relative humidity of 70-80% is best suited to the crop.
- Rice requires 30% of the full sunlight for its canopy to become light saturated. Low light intensity reduces total photosynthesis.
- During the growing period of rice, 17.2% of radiation is reflected, 30-47% is transmitted and 34.5 % is absorbed.
- Shading during initial growth period of rice causes reduction in effective tillers, length and grains.
- Rice is generally grown in the regions with rainfall of 1250mm. Its seasonal water requirement ranges from 350 to 700mm.
- There should be monthly rainfall of 200mm to grow lowland rice, and 100mm for upland rice. Land submergence upto milk stage is ideal for optimum grain yield.

8.3.4 Wheat

- In temperate regions, there are two cropping seasons of wheat. Winter wheat which matures in 10-11 months and spring wheat, which matures in 4-5 months.
- In India the wheat grown is spring wheat, which is grown during winter season.
- Wheat is the most important winter crop of north India, grown both as rainfed and irrigated crop. Cool weather during vegetative phase and warm weather during maturity is desirable for wheat.
- Wheat is a cold loving crop. The optimum temperature for seed germination is 20-25^{0}C, for growth of wheat optimum temperature is 15-20^{0}C with range of temperature from 10^{0}C to 25^{0}C.
- Higher temperature of about 30-35^{0}C have detrimental effects on growth. Higher temperature during winter accelerate growth, force early maturity reducing yield.
- Wheat crop is classified as long day crop requiring more than 12 h daylight during flowering period. Cloudy day mostly reduces grain yield.
- Decrease in day length period increases the length of growing period of wheat cultivars.
- As wheat is a C_3 crop, photosynthesis increases with increase in CO_2 concentration. Wheat crop converts 1.0 percent of solar energy to dry matter.

- Wheat is grown in the regions with rainfall 500-875 mm per growing season. Its moisture requirement is 450-650mm. Well distributed rainfall of 350-400 mm or four irrigations at critical growth stages produce optimum yield.
- Soil moisture stress at any stage of the crop decreases the grain yield. Soil moisture stress at crown root initiation stage significantly reduces tiller production. Water deficit at heading results in shriveled grains with poor test weight and low grain yield.

8.3.5 Maize

- Maize is the world's third most important crop after rice and wheat. Maize is best adapted to intermediate climates to which bulk of its cultivation is confined.
- In India, it is mainly grown during *kharif* season. It is also grown in summer as well as in winter season where irrigation facilities are available.
- It is grown in areas where mean temperature is more than 19°C and mean night temperature are more than 13°C.
- For germination, the minimum temperature required is 10°C and the optimum temperature in 30-35°C under adequate moisture.
- Optimum temperature for growth is 34°C. Higher temperature during reproductive phase reduces the grain yield.
- Improved cultivars are mostly day neutral and can be grown throughout the year provided temperature and moisture are not limiting.
- Maize is a C_4 plant. At low light intensity, the rate of photosynthesis is not affected by CO_2 concentration. At higher levels of light intensity, the rate of photosynthesis increases with increase in CO_2 concentration.
- Maize crop has high water requirement. It requires 500-800 mm rainfall during the season.
- It is resistant to drought and can recover from early season drought. Soil moisture stress during silking reduces the grain yield.

8.3.6 Sorghum

- Sorghum is well known for its capacity to tolerate conditions of limited moisture and to be productive during periods of extended drought.
- Sorghum is a C_4-type plant of tropical origin. It is productive at high light intensities and high temperatures.
- Sorghum is a heat loving crop. Because of its tropical origins, sorghum is very sensitive to low temperatures.
- The optimum temperature requirement for growth is 24-30°C within the temperature range of 15-35°C.

- Minimum temperature requirement for germination is 8-10 ^{0}C. Base temperature of 10^{0}C is taken to compute heat unit requirement of the sorghum.
- Most sorghums are sensitive to photoperiod and are classified as short-day plants, in other words, the night must be longer than a critical minimum.
- It is grown in the areas having low to medium well distrusted rainfall of 450 to 750 mm during the crop season. Sorghum cultivars with a life cycle of 110 to 130 days, require 450–650 mm of water.
- Sorghum is a drought resistant crop, becomes dormant during drought and returns to normalcy subsequently.

8.3.7 Pearl Millet

- Pearl millet is a cereal crop that is widely grown under rainfed conditions in the arid and semi-arid regions of Africa and southern Asia.
- Pearl millet, which accounts for about two thirds of India's millet production, is grown in the drier areas of the country.
- Pearl millet is suited to hot and dry climates, and can be grown in areas where rainfall is not sufficient (200–600 mm) for maize and sorghum.
- Millet production depends almost entirely on rainfall as its moisture supply. The amount and distribution of rainfall are important factors in determining the ultimate productivity of the crop.
- At sowing, poor soil moisture reduces seedling emergence, leading to poor crop establishment.
- During the vegetative growth period, the crop is well adapted to water deficits and can tolerate intermittent breaks in rainfall. During the early flowering and grain-filling stages, the crop is most sensitive to water deficits.
- During the germination and emergence stage, soil temperatures must reach 12°C for germination to begin. The germination rate increases linearly with temperature to a sharply defined optimum of 33°C and then drops sharply as temperatures increase.
- Pearl millet requires temperatures between 22°C and 36°C for a good photosynthetic response, with an optimum range of 31°C to 35°C and a lethal temperature around 45°C.
- Grain setting is optimum from 22°C to 25°C and declines at temperatures below and above this range.
- High temperature during flowering results in a loss of pollen viability and can reduce the receptivity of stigmas and affect grain filling.

- Day length, or photoperiod, is a critical control in the initiation of the reproductive phase of the millet in many pearl millet cultivars.
- Photosensitive cultivars are grown as long-season crops, while non-photoperiodic cultivars are grown as short season crops.
- Because of photoperiod sensitivity, the growth cycle of local millet cultivars changes greatly with sowing date.
- Pearl millet is a C_4-type plant, which means that it has a high photosynthetic efficiency, particularly under high temperature conditions, because of reduced photorespiration.

8.3.8 Groundnut

- Groundnut is an annual legume that is also known as peanut, earthnut, monkeynut and goobers. Groundnut is essentially a tropical plant and requires a long and warm growing season.
- Among oilseeds crops in India, groundnut accounts for about 50 per cent of planted area and 45 per cent of oil production.
- Although the crop can be grown in all seasons, it is grown mainly in the rainy season (*kharif*), running from June to September. The *kharif* season accounts for about 80 per cent of the total groundnut production.
- About 75 per cent of the groundnut area lies in a low to moderate rainfall zone with a short period of distribution (90–120 days).
- Minimum temperature for germination is 10°C, while the desirable is 20°C. Temperature in the range of 25°C to 30°C is optimum for growth and development. The temperature range for reproductive stages is 20-26°C.
- Groundnut is usually classified as day neutral plant. Pod yield is however, affected by day length. Long day (11-12 h) particularly at the beginning of the crop season, promotes vegetative growth and less reproductive growth.
- High humidity (95%) and low temperature (15°C) favour the incidence of leaf spot disease.
- The favourable climate for groundnut is well-distributed rainfall of at least 500 mm during the crop-growing season, accompanied by an abundance of sunshine and relatively warm temperature.
- Groundnut yields are reported to be vulnerable to the variation in rainfall. About 89 per cent of yield variation in India was attributed to rainfall variability during August to December growing period.
- Pod yield of rainfed groundnut depend on the distribution of rainfall during August and September. Soil moisture stress at different stage causes variation in duration of the groundnut.

8.3.9 Soybean

- Soybean is a tropical crop but its cultivation now extends to subtropics and temperate climates.
- Germination is the most critical stage in soybean and requires minimum temperature of 8-10^{0}C for germination.
- The optimum temperature for soybean growth is 24-30 ^{0}C. The lowest temperature for most growth processes is taken as 10^{0}C.
- High night temperature increases vegetative growth. Low temperature (<20^{0}C) delay the flowering. High temperature (>33^{0}C) causes reduction in yield.
- Soybean is basically a short day plant. Increased day length may result in the delay of flowering. Critical photoperiod for flower initiation is 14 h.
- Soybean is mostly grown during *kharif* season in the regions having seasonal rainfall of 550-750mm. It can also be grown during rabi season with supplemental irrigations.
- Soil moisture stress at flowering and pod development stage reduces the seed yield by 20-50%.

8.3.10 Mustard

- Rapeseed and mustard are important oilseed crops of India, next to groundnut. These are grown as rainfed crops in *rabi* on residual soil moisture after *kharif* crops.
- Rapeseed and mustard are crops of tropical as well as temperate zones and require somewhat cool and dry weather for satisfactory growth.
- Crop requires about 18-25°C temperature, low humidity, practically no rains especially at the time of flowering. The crop growth is optimum at <25°C, while it ceases at less than 3°C and >35°C.
- Mustard is classified as a long day plant. The long duration mustard is usually grown as an irrigated crop on relatively light soils.
- Mustard generally requires fair supply of moisture, restively dry weather and plentiful sunlight during growing season.
- It is sensitive to frost, at seed development stage reduces the seed yield.
- Heavy wind at pod maturity leads to shattering and lodging of plant leading to poor yield.
- High humidity and cloudy weather during pod development is not desirable as it results in pest and disease incidence.

8.3.11 Cotton

- Cotton is the world's most important fibre crop and the second most important oilseed crop. Cotton is grown in over 60 countries around the world.
- Cotton is a heat loving plant and best adapted to subtropical climates. Generally, cotton requires abundant sunshine, adequate moisture and fairly high temperature.
- The optimum temperature requirement of cotton for the growth and development is 15-20°C with lower and upper limits of 10°C and 25°C respectively.
- Cotton requires a minimum daily air temperature of 15°C for germination, 21°C–27°C for vegetative growth and 27°C–32°C during the fruiting period..
- The cardinal temperatures for cotton growth are 15-20°C, 25-30°C and 35-40°C.
- Higher temperature (>35°C) has detrimental effect on rate of growth and fibre production.
- Adequate soil temperature and moisture conditions at planting are necessary to ensure proper seed germination and crop emergence.
- At least 500 mm of water (rainfall and/or irrigation) is required to produce a cotton crop. Cotton needs between 550 mm and 950 mm rainfall during the season in a consistent and regular pattern.
- Environmental conditions such as overcast skies, rainy weather, water deficits and high temperatures will decrease photosynthesis and the supply of photosynthate.
- Water stress caused by a deficiency of water manifests its damage as reductions in photosynthetic activity and increases in leaf senescence.
- Moisture deficit stress reduces plant growth, resulting in stunted plants with reduced leaf area expansion. Drought stress causes severe shedding of small squares, resulting in a decrease in flowering
- Wind can also stress the cotton plant enough to reduce yield. Wind modifies the temperature and humidity gradients around the cotton plant, which in turn changes the evaporative demand.

8.3.12 Sugarcane

- Sugarcane is a tropical plant and thrives best in hot sunny area where temperature, light and moisture are principal climatic factors that controls the cane yield.
- Arm summer growing season with adequate rainfall/irrigation is an ideal climate for sugarcane production.
- Minimum soil temperature required for germination is 19-21°C, while the optimum is 27-28 °C.

- Air temperature <21°C checks the vegetative growth, induces ripening and leads to better juice quality.
- Tillering is maximum at 30°C. During tillering the minimum temperature <18°C causes decrease in cane yield.
- Day temperature of 27-31 °C and night temperature of 22-23°C is conducive for flowering.
- Sugarcane is a sun loving plant. Therefore, greater incident radiation (Sunshine) favours higher sugarcane and sugar yields. Flower initiation occurs when the photoperiod is 12h.
- In India sugarcane is grown in areas ranging in rainfall of about 600 mm to 3000 mm.
- The crop water requirement of sugarcane is 1500-2000 mm in subtropical region and 2000 to 3000 mm in tropical regions of the country.
- Pre-monsoon is critical period for rainfall/irrigation. During post monsoon season, relatively low soil moisture facilitates ripening and increases the sucrose content.

8.3.13 Potato

- The potato is a major world food crop and by far the most important vegetable crop in terms of quantities produced and consumed worldwide.
- Climatic conditions, as affected not only by the latitude but also by altitude, influence potato plant growth and development.
- Potato is best adapted to cool climates such as tropical highlands with mean daily temperatures between 15°C and 18°C. There are three phenological phases of Potato, all are influenced by climatic factors differently.
- The duration of the first phase, comprising the development period between emergence and tuber initiation, is shortened by short days and temperatures less than 20°C.
- The duration of the second phase is affected by temperature, with an optimum between 16°C and 18°C and by solar radiation. Crop senescence is shortened by high temperatures, especially greater than 30°C.
- Low mean temperatures (15°C–19°C) with a short photoperiod (12 h) were most suitable for early tuber growth.
- Tuber development declines as soil temperatures rise above 20°C and tuber growth practically stops at soil temperatures above 30°C.
- An adequate water supply is required from tuber initiation up until near maturity for high yield and good quality.

8.4 CLIMATE RELATIONS WITH LIVESTOCK AND FISHERIES

8.4.1 Livestock

- All animals have a thermal comfort zone, which is a range of ambient environmental temperatures that are beneficial to physiological functions.
- During the day, livestock keep a body temperature within a range of ±0.5 °C. When temperature increases more than the upper critical temperature of the range (varies by species type), the animals begin to suffer heat stress.
- Animals have developed a phenotypic response to a single source of stress such as heat called acclimation. Acclimation results in reduced feed intake, increased water intake, and altered physiological functions such as reproductive and productive efficiency and a change in respiration rate.
- Heat stress on livestock is dependent on temperature, humidity, species, genetic potential, life stage, and nutritional status.
- Livestock in higher latitudes will be more affected by the increase of temperatures than livestock located in lower latitudes, because livestock in lower latitudes are usually better adapted to high temperatures and droughts.
- Confined livestock production systems that have more control over climate exposure will be less affected by climate change.
- Heat stress decreases forage intake, milk production, the efficiency of feed conversion, and performance.
- Warm and humid conditions cause heat stress, which affects behavior and metabolic variations on livestock or even mortality. Heat stress impacts on livestock can be categorized into feed nutrient utilization, feed intake, animal production, reproduction, health, and mortality. The following presents these in more detail.
- Failure to meet the dietary needs of cattle during heat stress affects metabolic and digestive functions.
- Thermal livestock stress decreases feed intake and efficiency of feed conversion, especially for livestock that are fed large amounts of high quality feeds.
- One of the major causes of decreased production in the dairy and beef industry is heat stress. High-producing dairy cows are more sensitive to heat stress.
- The poultry industry may also be compromised by low production at temperatures higher than 30°C. Heat stress on birds will reduce body weight gain, feed intake and carcass weight, and protein and muscle calorie content. Heat stress on hens will reduce reproduction efficiency and consequently egg production because of reduced feed intake and interruption of ovulation.

8.4.2 Fisheries

- Climate-related changes that affect ecological functions and the frequency, intensity and location of extreme weather events include: changes in salinity and freshwater content; oxygen concentration; carbon uptake and acidification; temperature and thermal stratification; sea levels; ocean circulation; surface wind, storm systems and waves.
- These changes can be expected to have a range of impacts, both direct and indirect, on fisheries and aquaculture.
- Evidence exists that climate change is modifying the distribution of marine species. Many species are migrating towards the poles and deeper waters to follow their ideal habitat conditions.
- Rising sea levels can displace brackish and freshwater systems in delta zones and wipe out a range of agricultural practices.
- The impacts of climate change on freshwater fisheries and aquaculture are expected to be significant.
- The increased variability in levels of precipitation and changes in air and water temperatures will affect the productivity of rivers, lakes and floodplains.
- Climatic drivers (e.g. higher temperatures) affecting inland ecosystems and species distribution are often intensified by non-climatic drivers, such as invasive species, pollution and habitat modification, and the fragmentation of rivers by dams.
- At the regional level, freshwater reservoirs will also be increasingly under pressure to meet the growing demand for irrigation.
- In general, inland fisheries will be at risk in areas where water stress is acute and the competition for water resources is high.

9

ANIMAL CLIMATOLOGY

9.1 STRESS ON ANIMAL

9.1.1. Environmental stress

- Heat stress
- Cold stress
- High humidity

a) Heat stress-

It causes

- Loss of production
- Loss of body gain or body weight
- Poor reproduction
- Rapid respiration (it increases from 40-50% in summer in buffalo)
- Flank region used for counting of respiration
- Increase in body temperature

Normal temperature Animals/Birds

100.5-101.5°F Cow, buffalo

107°F Poultry

101-102°F Pig

- rapid pulse rate

55-65/min is normal in cows and buffaloes

70-80/min is abnormal or fever condition

- Buffaloes are known as seasonal breeders because they came to heat in only rainy and winter season.
- Impaction of rumen due to feeding of dry fodders without any water so the cow dung become hard to come.

Control measure of heat stress:

Not allowing the animal for pasture grazing in hot summer time (outside).

Providing shade-

- Natural shade (trees)
- Artificial shades (barns- kachcha and pakka types)

Cooling systems

- Fans + exhaust fans
- Coolers
- Fans with mist
- A/C
- Proper ventilation

Wallowing (pakka tank with proper drainage)

Bathing of animals (clean milk production)

b) Cold stress or chill stress

- Body temperature may increase due to disease condition when proper care is not given during winter.
- During winter very chill, foggy weather, rain with wind causes pneumonia.

Control measures

Heat is provided by fire using traditional materials.

- Room heaters
- Use of curtains
- Bulbs

Covering the body of animals using old gunny bags or blankets.

Balanced diet should be provided (protein content should be more). Like oilseed cake, mustard cake, til cake etc.

Stress of humidity:

- More sweating (body minerals decreases)
- Serious weakness

- It is more dangerous than heat and chill stress because due to this animal loses more body minerals and sweat more.
- More fans are required with more exhaust systems.

2. Transportation stress-

Due to tiredness;

So, there are many transportation shelters in which animals are maintained properly with proper care in order to avoid transportation stress.

- Animal rest
- Proper grooming and massaging.

3. Vaccination and medicinal stress-

- Loss of appetite (which causes loss of production)
- Production loss
- Reproduction problem

Control measures-

- Avoid excess dose
- Vaccination should be done properly

4. Excitement or nervous stress-

It occurs very occasionally, may be due to animals like cats, dogs etc.

Control measures of stress-

1. Always provide pleasant environment to the animals.
2. Various medicines/use of antistress medicines.

a) Electrocare (minerals) for animals-

- 100-200gm for adult animal.
- 100gm per 1000 chicken with water.
- 50-100gm per young animals.

b) Zee tress (herbal drug)-

- Any stress in chicken can be cured by it.
- 5-10gm/adult animal daily
- 1-2gm/young animal
- 5gm/1000 chickens

c) Stress rock (medicines)

- 10-15ml/1000 chickens
- 20-25ml/animal

d) Stress cool (it may be in liquid form or tablet form)

- 20-25ml/animal
- 10ml/1000 chickens

Care of pregnant cows

1. Before cowing

- Keep the advance pregnant cow away from the herd.
- Provide them shade or barn.

(Cowing box or cowing pan- specific place for pregnant cow for proper care before and after cowing).

- Provide easily digestible diet and sufficient amount of fresh and clean drinking water.
- Protect the cow from wild animals, street dogs and wild birds.
- Daily cleaning and disinfection of barns.
- Avoid use of slippery floors.

2. During cowing

- Take care of animals during cowing as mentioned above.
- Make proper arrangement of one attendant to look after the cow during cowing.

3. After cowing

If there is the case of retention of placenta, remove it after six hours by taking the help of veterinarian.

Effect of heat stress on production

In most tropical feeding systems, the forage is having low nutritional values during summer as compared to the winter. Initially the feed intake is depressed (decreased) under hot condition and the efficiency of nutrient metabolism and absorption of nutrients is reduced. So low nutritional quality and high environmental temperatures have various effects on reproduction. Heat stressed cows are likely to exhibit heat and often should heat at night when they are less likely to be observed. In addition, duration of estrus is shorter for cows subjected to heat stress. The time return to the activity is closely linked to blood glucose level of the cows, indicating the effect of increased heat load of feed intake and energy metabolism are very important. Increase in heat loads directly reduces utilizing hormone secretion. There are several reasons which affects reproductive performance during hot weather. Heat stress alters the follicular development pattern in cattle.

Higher intrauterine temperatures likely reduce embryo survival and also have adverse impact on consumption. One reason for greater temperature of uterus is the higher body temperature and less uterine blood flow to remove heat, because more blood is shifted to the skin to help reducing body heat.

Management strategies to optimize/minimize reproductive heat stress:

There are specific steps that can be taken to mitigate effect of heat stress on reproduction in dairy cattle.

1. Shade

All animals including dry cows and pregnant cattle should have access to effective shade during the hot period of the year.

2. Cooling

In severe heat stress provision of stress is some for cooling may enable the cow to maintain their normal body temperature.

*Dry cows:

The cows who stops milking and in advance pregnancy (dry period 45-60 days).

*Heat cycle:

For 21 days; cow remain in heat period for 16-22 hours.

*Lactation (milking) period- 300 days

*Gestation period- cows-280±5 days

Buffalo- 310 days

*Herd average= (milking cow + dry cow)/no. of total cows

*Milking average= Milking cow/ total cows

Adaptability

It is the ability of animal to withstand against the adverse conditions and adjusts themselves to environment in which they live.

Adaptation

It is the process by which an animal adopts itself into new changed environmental condition. It is of 3 types as follows-

1. Biological adaptation

- Morphological
- Anatomical

- Biochemical
- Behavioural

Characteristics of the animal promoting its welfare and favouring its survival in a specific environment.

2. Genetic adaptation

The changes which taking place over many generations of population in a particular environment which occurs to genetic adaptation.

3. Physiological adaptation

It involves capacity and process of adjustment of animal to the external physical environment.

Adaptability can be evaluated by the animal capability to adjust environmental conditions and climatic extremes.

Well adapted animals are characterized by-

- Minimum loss in body weight when exposed to stresses like nutritional deficiency, transportation etc.
- Highly resistance to disease.
- Low mortality rate.

Acclimatization

It is the sum of all long term complex physiological process by which animal adjust to environmental conditions.

Loss of water from the body-

- Dehydration
- Sweating
- Urine
- Major factor responsible is diarrhea.

Control

- Fresh water- adlibtum (as much as they can take)
- Jaggery water
- Rintose- 25% glucose (provide glucose to animal during water stress condition)
- Calcium supplements
- Mifex solution for energy.

Ambient Temperature			Humidity with temperature		
Parameters	**Low**	**High**	**Low**	**High**	**Radiation**
1	2	3	4	5	6
A. Physiological					
1. Rectal temperature	Normal	Rise	Normal	Increase	Increase
2. Pulse rate	Normal	Increase	Normal	Increase	Increase
3. Body heat load	Less	More	Less	Increase	Increase
4. Muscular heat production	More	Decline	More	Decrease	Decrease
5. Chloride content	Normal	Increase	Normal	Increase	Increase
B. Behavioural					
1. Appetite	Normal or Decline	No effect	Decline	Decrease	
2. Food intake	Slightly or more	Decline	Normal	Decrease	Decline
3. Water intake	Decrease	Increase	Low	Increase	Increase
4. Seeking more shade	No	Yes	No	Yes	Yes
5. Sluggishness	No	Yes	No	Yes	Yes
6. Browsing more at night	No	Yes	No	Yes	Yes
7. Grazing time	Increase	Decrease	No effect	Decrease	Decline
8. Respiration rate	Normal	Increase	Normal	Increase	Increase
C. Productive					
1. Growth	Better	Decline	Normal	Decrease	Decline
2. Milk yield	Increase	Decline	Normal	Decline	Decrease
3. S.N.F	Normal	Decline	Normal	Decline	Decline
4. Fat yield	Normal or increase	Decline	Normal	Decline	Decline
5. Lactose & nitrogen content	Normal	Decline	No effect	Decline	Decline
D. Reproductive					
1. Breeding efficiency	Better	Decline	No effect	Decrease	Decrease
2. Spermatogenesis	Normal	Decrease	Normal	Decline	Decrease
E. Others					
1. Body weight	Normal	Decrease	No effect	Decline	Decrease
2. Body size	Higher	Decline	No effect	Decrease	Decline
3. Belly girth	No effect	More	No effect	More	More
4. Skin burn etc.	No effect	Yes	No effect	Yes	More

10

APPLICATIONS OF METEOROLOGY TO AGRICULTURE

10.1 INTRODUCTION: AN OVERVIEW

The application of meteorology to agriculture is essential, since every facet of agricultural activity depends on the weather. This chapter discusses many examples of the usefulness of applied meteorology in order to make users aware of the potential benefits that farmers can gain to improve efficiency and ensure sustainability of their farm management; to protect and ensure the continuing health of their crops, livestock and environment; to increase their yield and the market value of their crops; and to solve selected operational problems. Other agricultural decision-makers derive benefit from agrometeorological applications, including government policymakers seeking to ensure adequate food supplies, affordable food prices for consumers and sufficient farm income for farmers, and to reduce the impact of agricultural practices on the environment. Decision-makers in international agricultural organizations also use applications of meteorology to ensure food security and to react to potential famine situations. At any level, these objectives can be achieved only through active cooperation among National Meteorological and Hydrological Services (NMHSs), agricultural extension services, farmers and their associations, agricultural research institutes, universities and industry. The topics discussed in this chapter are based on a classification of applications of micrometeorology to various agricultural problems. They are grouped according to the general type of problem, namely, improving production; averting dangers to production; physiology and growth; strategy; and tactics. In selecting the examples of applications of meteorology to agriculture, priority has been given to those that are used operationally. Also, an attempt has been made to cite examples for the different climates and regions of the world.

10.1.1 Users of Agrometeorological Information:

The successful application of weather and climate information needs to integrate three components: data, analysis and users. Therefore, the ultimate goal of any application is to serve the needs of the users. A solid foundation of data is a prerequisite for successful agricultural meteorological applications. Then an analysis of these data that seeks to solve or address an agricultural problem is needed. These analyses are described in this Guide. Ultimately, the users of agrometeorological information and applications must be kept in mind when developing new applications. The user can be defined as any agricultural decisionmaker such as a farmer, extension agent, government official, media person or the general public

10.1.2 Temporal and Spatial Scales of Agrometeorological Information and Applications

Agrometeorological information and applications can be considered in temporal and spatial contexts. In a temporal context, strategic applications are defined as those aiding in issues and decisions that are assessed on a seasonal or yearly basis or only once, such as in a planning process. These applications aid in the planning process whether the decision is choosing a specific crop variety to plant, if an area should be exploited for forage products and livestock, how to design and plan where or if greenhouses or animal shelters should be built, or how to assist governments in setting agricultural pricing policies. Such decisions can be based on climatological analyses, agroclimatic information, and the use of complex soil–plant–atmospheric models. Tactical applications are considered to be short-term operational decisions relating to a period ranging from a few hours to a few days. These often involve decisions, based on the state of the crop and current or forecast weather, for such farm operations as cultivating, irrigating, spraying and harvesting.

Agrometeorological Aspects of Organic Agriculture, Urban Agriculture, Indoor Agriculture and Precision Agriculture provides a good description of macroclimate and mesoclimate in the context of agrometeorological applications. Macroclimate is the largest and covers broad areas of a continent (millions of square kilometres), and deals with the interaction of large-scale topography (mountain ranges, large lakes and ocean influences) with airmasses. At this scale, climate characteristics should provide information on the suitability of a farm and whether the farm could be weather-limited by pest, disease and operational timing problems. Mesoclimate reflects the farmer's view of the weather experienced in a region. Local surface features such as hills, small mountains, large forests or extensive plains have a distinct effect at this scale. A country may have one or two macroclimate zones, but it will have many mesoclimates. It is at this scale that specific calculations can be made to define agroclimatic regions.

At the smallest scale, microclimate is defined as the climate near the ground, or in other words, the climate in which most plants and animals live. In terms of meteorology, The Application of Micrometeorology to Agricultural Problems describes micrometeorology as dealing with the physical processes taking place within the boundary layers between the top of the plant, tree or animal and the bottom of the roots of the soil. Most of the applications in this chapter are based on micrometeorological principles.

10.1.3 Benefits Derived from Applications

Many benefits result from the application of meteorological services to agriculture. The productivity of a region or of a particular enterprise may be increased by the reduction of many kinds of loss resulting from unfavourable climate and weather, and also by the more rational use of labour and equipment. Greater economy of effort is achieved on the farm, largely through a reduction in activities that have little value or are potentially harmful. All of these increase the competitiveness of production, reduce risk and help to lower the cost of the final products. In the developed world, a significant portion of recent work in agricultural meteorology has shifted from increasing yields to reducing the environmental impact of agricultural fertilizer and pesticide use and combating pests and diseases. In the developing world, much of the focus remains on increasing agricultural production, but there is also an emphasis on sustainable agricultural production and reducing the impact of diseases and pests such as desert locusts. In Sudan, precise calculations of water requirements for the main irrigated crops (cotton, sorghum and groundnut) were compared with available irrigation water to allow for more accurate estimates of potential irrigated wheat area. The net result was an additional 8 000 ha of wheat grown, which added more than US$ 2 million to the national economy at a cost of a few thousand dollars for data, analysis and staff. In the Gambia, groundnuts are stored in the open air and if the dry pods become wet, they are at high risk of developing aflatoxin, which can reduce farmer prices for the crop by up to 60 per cent. If farmers are warned of rainfall by forecasts transmitted via local radio, they can cover the crop with plastic sheeting. It is estimated that for each percentage point of production saved, the benefit is US$ 60 000. In the Sahel, bush fires are common every year, but the bush vegetation is needed for cattle and sheep grazing. By using wind, temperature and humidity observations to indicate speed and direction of the fire, controlled burning can take place to prevent the fires from spreading. Reducing the burned area on 1 per cent of the grazing land allows 5 000 more sheep to graze, which represents an additional annual value of US$ 100 000 to the national economy. A WMO report on meteorology and plant protection provides a framework for analyzing costs and benefits of agrometeorological applications in plant protection The results of studying the peculiarities of climate and weather conditions to optimize various cultural practices (namely, determination of an advisable structure of areas under crops, dates and methods of soil treatment, optimum fertilizer application periods and doses) aimed at boosting the productivity of plants growing in Russia. The economic efficiency of applied agrometeorological recommendations is given. Other examples can be found in The Economic Value of Agrometeorological Information and Advice and in materials from a conference on the economic benefits of meteorological and hydrological services.

10.2 APPLICATIONS FOR GOVERNMENTS AND OTHER LARGE ADMINISTRATIVE BODIES

Governments and other large administrative bodies need high-quality and reliable information for operational assessments of agricultural production. With regard to planning, this would involve questions about what kind of crops the country could produce economically and where they could be grown. Planning questions of this nature can be answered by macro- and mesoscale agroclimatic surveys.

10.2.1 Operational Assessments

There are several examples of operational assessments of crop production that countries and international agencies perform. These examples highlight the utility of integrating data, staff and resources to produce reliable crop production assessments among different agencies at a country level and among different countries, non-governmental organizations (NGOs), and other organizations at an international level, as well as the need to do so. Researchers provide many examples of successful drought monitoring and crop monitoring applications at these levels. The Food and Agriculture Organization of the United Nations (FAO) established the Global Information and Early Warning System (GIEWS) in 1975 in response to the global food crisis of the early 1970s. GIEWS provides information on food production and food security for every country based on crop monitoring assessments that involve remote-sensing and ground-based weather station data. A number of these assessments are based on the FAO Crop Water Requirement Satisfaction Index (WRSI) model that determines a cumulative water balance for 10-day periods from planting to maturity. The WRSI model is a combination of dynamic water balance and statistical approaches and the index represents at any time of the growing season the ratio between the actual and potential evaporation.

The World Agricultural Outlook Board (WAOB) of the United States Department of Agriculture (USDA) is mandated to provide official monthly United States Government forecasts of agricultural commodities through the World Agricultural Supply and Demand Estimates (WASDE) publication. These supply and demand estimates are based on official country reports, United States embassy reports, travel reports of USDA personnel, economic analysis, remote-sensing information and, of course, global weather information. USDA meteorologists routinely collect, monitor and analyse global weather conditions and agricultural information to determine the impact of growing-season weather conditions on crops and livestock production prospects. These activities are supported by meteorologists from the United States National Weather Service who are located within WAOB offices to serve the agricultural community.

The application of an automated Russian forecasting system that allows for the presentation of information about observed weather conditions, state of crops, crop yield forecast and the total harvest in the various regions of Russia. Comprehensive application of climatic and weather information at the governmental and field levels is demonstrated with the Russian information and advice system meant for resolving some practical problems in the planning and organization of agricultural production.

In Brazil, several centres generate daily updated agrometeorological information to support decision making, including Agritempo (http://www. agritempo.gov.br), which provides information for the whole country, and regional centres such as Cepagri/ Unicamp (http://www.cpa.unicamp.br) and IAC (http://www.ciiagro.iac.sp.gov.br) for the state of Sao Paulo, IAPAR/SIMEPAR (http://www.iapar.br/ sma) for the state of Parana, and Ciram (http://www. chlimerh.rct.sc.br) for the state of Santa Catarina. Farmers' cooperatives also provide agrometeorological information to their farmers and field technicians (for example, http://www.fundacaoabc.org.br) (Pinto, personal communication).

Other examples include the Famine Early Warning System, or FEWS, the European Union's Monitoring Agriculture with Remote Sensing (MARS), the National Agricultural Monitoring System (NAMS) in Australia, the Farm weather service and agrometeorological services in Kazakhstan.

10.2.2 Agroclimatic Survey

Concerning macroscale agroclimatic surveys, the joint inter-agency project on agroclimatology sponsored by FAO, WMO and the United Nations Educational, Scientific and Cultural Organization (UNESCO) has been very successful in producing five publications and is the foundation of current agroclimatic zoning studies. The most important practical applications of macroclimatic surveys include: choosing crops, varieties and domestic animals; determining favourable periods for sowing, haymaking and harvesting; establishing areas where dryland farming is possible and where irrigation has to be applied; planning afforestation and reforestation; finding the optimum range of climatic variables for increasing yields and agricultural production in general; establishing potentials for the agricultural use of rangelands; and determining requirements and potentialities for efficient storage and transportation of crops.

The agroclimatology in the semi-arid and arid zones of the Near East has focused on the estimation of the boundary areas where dryland farming is possible and where irrigation is needed. The agroclimatology of semi-arid areas in West Africa has dealt with dryland farming in West Africa and the length of growing seasons, which is strictly associated the rainy season. The agroclimatology of the highlands of eastern Africa has focused on which crop water requirements are met in the various localities of this area. The agroclimatology of the Andes has looked at the unique effects of high elevation and high solar radiation input on crops. An agroclimatology survey of the humid tropics of South-East Asia illustrates the role of agroclimatology in determining strategies to increase food production in the humid tropics.

10.2.3 Mesoscale Agroclimatic Surveys

With the advent of widespread computing capability (personal computers) and Geographical Information Systems (GISs), climatic and agroclimatic mapping has become widespread. There are many examples of the use of these methods for climatological and agroclimatic analyses. The European Union-sponsored European Cooperation in Science and Technology (COST) Action 719 on the Use of Geographic Information Systems in Climatology and Meteorology provides several examples of these agroclimatic mapping methods, including temperature mapping, climate parameter mapping in mountainous areas, and an agroecological decision system. The Parameter–Elevation Regressions on Independent Slopes Model (PRISM) has been used to map daily weather and climatic parameters in mountainous areas.

Scientist focused on the development of a methodology to assess climatic and agroclimatic risks. Land capability was classified for Sardinia by using climate, geographic and soil data. A climatic risk index was computed on the basis of 30-year averages of climatic data. A mesoscale agroclimatic classification scheme for Czechoslovakia based on a hydrothermic coefficient, HTC = R/(0.1TS10) where R is the rainfall sum in millimetres and TS10 is the degree-days above 10°C. Such an approach could be used with any degree-day base temperature and adapted for specific crops.

In Brazil, there have been recent nationwide efforts in agroclimatic risk zoning for agricultural crops that characterizes the potential and climatic risks for several crops, including maize, soybeans, beans, wheat, barley, rice, cotton, coffee, cassava and different species of fruits (Pinto, personal communication). The recommendations are used by the Brazilian Government to provide financing to the farmers at very low rates. Furthermore, those who follow the recommendation of the agricultural zoning can obtain official insurance at special rates. To be eligible for the bank credit, the farmers must also adopt the best agronomical practices recommended by the extension service. These efforts, combined with farmers' use of optimum planting dates, have increased the productivity of Brazilian agriculture (Pinto, personal communication).

10.3 APPLICATIONS FOR FARMERS OR GROUPS OF FARMERS

10.3.1 Improvements to Production

- **Irrigation** In its broadest terms, irrigation involves water balance calculations based on rainfall, estimation of water infiltration (effective rainfall), runoff, evapotranspiration (ET) and soil moisture. There are several reliable direct measurements for soil moisture, such as those obtained using manual gravimetric and neutron probe methods, which are suitable for routine application in agricultural practice. Indirect measurements based on remotely sensed information are also possible.
- Smith (2000) provides a survey of the widely accepted practical procedures that have been developed by FAO et al. to estimate crop water requirements and yield response to water stress. A review and update of the methodologies are contained in FAO Irrigation and Drainage Paper No. 56, which deals with crop evaporation. These methodologies use the Penman–Monteith equation, which estimates daily reference crop evapotranspiration (mm/day) based on net radiation, soil heat flux,

average air temperature, wind speed, vapour pressure deficit, and other humidity parameters. The two publications listed above give details on estimating all these parameters based on weather and climate data and when data sources are limited. The FAO CROPWAT software program incorporates these methodologies and procedures to simulate crop water use under various climate, crop and soil conditions. A report published by the WMO Commission for Agricultural Meteorology describes several operational applications to increase water use efficiency, including an irrigation advisory system in Israel calculated on the basis of a modified Penman potential ET equation. The same report contains a paper describing the Irrigation Planner, which has been developed into a computer software application for irrigation of grassland in the Netherlands. Results show that using the system can reduce irrigation water by 15–20 per cent.

- **Shelter from the wind** Rosenberg et al. (1983) define windbreaks as structures that reduce wind speed, and shelterbelts as rows of trees planted for wind protection. Both of these can reduce physiological stresses on plants and animals due to wind. Rosenberg and his co-authors reviewed the literature and found that shelter effects on the microclimate include reduced potential and actual evapotranspiration; improved internal plant water relations (greater internal water potential and lower stomatal resistance); improved opportunity for photosynthesis; and finally, a general increase in yield as a result of shelter. These generalities are subject to variation depending on soil moisture, and the benefits may be most dramatic in dry years or under critical moisture shortages. Examples of the widespread use of windbreaks can be seen in the Great Plains in the United States after the Dust Bowl years of the 1930s, the Rhone Valley in south-eastern France and the Netherlands. Marshall (1967) has reviewed the literature on the effect of shelter on the productivity of grasslands and field crops, and showed how the proportional decrease in wind speed with distance from the shelter corresponds to a decrease in evaporation. Night-time temperature decrease, relative humidity and the increase in daytime air and soil temperatures vary with distance from the barrier, but decline to no effect at a distance of about 12 times the height of the shelter. In connection with these parameters, the greatest soil moisture availability and crop yield are found in the zone at a distance of 2 to 4 times the height of the shelter. Windbreaks reduce the force of the wind in the sheltered zone. It shows that a dense barrier may protect an area about 10–15 times the height downwind, and by increasing the porosity of the barrier to about 50 per cent, the downwind influence can be increased to 20–25 times the height. Rosenberg et al. (1983) state that for the best wind reduction and greatest downwind influence, the windbreak should be most porous near the ground and the density of the barrier should increase logarithmically with height in accordance with wind speed profile. Wind reduction is a function of shelter location as well as the height above the plants.

Questions of orientation and spacing of shelter can be regarded as meteorological applications, particularly if mesoclimatic wind surveys are used in advance.

- **Shade** Shelters of various types can also be used to provide shade from the sun; a well-known example is the use of taller-growing "shade trees" to protect cacao, coffee or tea plants. Agrometeorology of the Coffee Crop states that because coffee originally developed as an understorey shrub in the rain forests of central Africa, it might be assumed that shading or arborization of coffee trees is a well-defined cultural practice. There has been much discussion on the validity of this practice, however. Most of the commercial crop in Brazil is unshaded, while shading is a common practice in Colombia, Costa Rica, El Salvador, Guatemala, Uganda, Tanzania and in the higher-elevation areas of North-east Brazil. In most of Brazil, an unshaded crop facilitates the harvesting and natural ground drying of the crop as the microclimate is sunnier, and hence warmer and drier. In most places, coffee berries are hand-picked at the cherry stage, and if this stage is extended it provides for an easier and longer harvest period. Therefore, shading can be advantageous since it increases the cherry stage of the crop because the microclimate is cooler and moist, which slows the maturation process. Shading also aids in maintaining high soil organic matter. Shelters may also be used to reduce production losses from lactating dairy cows because of the heat load during the summer.
- **Greenhouses (glass and plastic)** Greenhouses have been used in temperate climates for over 100 years and serve mainly to reduce heat loss and permit complete control over the watering of plants. Recently, CO2 enrichment of the atmosphere has become an additional technique in greenhouse cultivation. It discusses many benefits derived from indoor agriculture (greenhouses): protection against damage by ultraviolet (UV) light; improved ambient temperature conditions; protection of crops from adverse climatic conditions; increased productivity; reduced production costs; controllable harvest; and better product quality. They also list several climate elements that must be managed for good performance from greenhouses. The covering of the greenhouse is important with regard to visible light transmission for plant photosynthesis. Knowledge of the climatology of a greenhouse site, including solar radiation, cloudiness, relative humidity, temperature and wind profiles, is important.
- Turning to construction materials, most greenhouse coverings are made of glass, fibreglass and plastic, while plastic agricultural tunnels are less widely used. In order to choose the best covering suited to a geographical location, the maximum, minimum and average temperatures; the possibility of frost; the climatology of the wind and relative humidity, rainfall distribution and intensity; solar radiation; and specific crops need to be taken into account. A climatological analysis of solar radiation, temperature and relative humidity is important for siting greenhouses.

These parameters will determine how much internal environmental control will be needed for optimum plant growth depending on the plants grown. Wind speed and direction are very important factors when designing a greenhouse. High winds could damage the structure or coverings. Wind is also used in simple greenhouse designs to maintain the thermal balance by reducing energy costs for heating or cooling. Wind ventilation can be used for balancing internal temperatures by means of air circulation, reducing relative humidity, promoting crop pollination, and replenishing carbon dioxide and removing oxygen for plant photosynthesis. For some crops the degree of control is such that firm advice can be given on the optimum temperatures for different growth stages (tomatoes are a good example); it is possible to differentiate between the environmental temperatures that should be maintained during the day and during the night. This knowledge has led to the design of "blueprints" for the production of certain crops. There is scope for further assistance by meteorologists in research on environmental control. Practical help can also be given at the advisory level, in terms of greenhouse siting, design, and fuel consumption. Meteorological factors are probably most useful in siting greenhouses. In analysing possible greenhouse sites, standard radiation data can be adjusted for latitude and mean cloudiness to give an estimate of the radiation input (and therefore plant growth) at each location. Such factors as shelter and radiation must be balanced; highly exposed sites are undesirable because of extra fuel consumption and the risk of physical damage. As for fuel consumption, if a crop requires the temperature to be kept at a given level, the quantity of fuel needed can be calculated from the number of degree-days below that temperature.

- **Ground cover (mulching)** Soil mulches of various kinds (namely, straw cover or artificial materials such as plastics) are used to modify the heat and moisture balance in the soil to benefit plants. It provides a good overview of the effects of mulching on plant climate and yield and states that mulches are particularly useful in conserving moisture, reducing temperature extremes and minimizing erosion. Agrometeorology of the Coffee Crop lists the following advantages of grass-straw mulching with the coffee crop in Kenya: it protects the soil from excessive heating that destroys the soil structure; it lowers temperatures, which results in lower evaporation rates; it provides organic matter to the soil; and it reduces soil erosion from heavy rainfalls and minimizes weeds. It also cites studies indicating that mulching can reduce the frequency of irrigation. On the negative side, mulching requires a large amount of grass, and more importantly, straw mulches can aggravate frost problems as the air temperature above the mulch is much warmer during the day and cooler at night. The mulch also prevents the ground from absorbing heat during the day, which is subsequently released during the night. Studies on the effect of straw mulching on air temperatures indicate that at 5 cm above the ground, the maximum temperature is 6.6°C higher and the minimum is 1.7°C lower than the bare ground. On areas subject to frost, transparent plastic should be used and white plastic should be used elsewhere, since it approximately has the same thermal regime as bare soil.

- **Animal housing** Meteorological data are required when assessing whether and how animal housing should be put into use. Evaluation should also take into account the potential economic returns, energy cost and availability. Animal housing is utilized because thermal imbalances lead to adverse effects on animal productivity. Weather and climate can determine the efficiency of livestock production by direct and indirect influences. Direct influences affect the heat balance of the animal and include extreme meteorological events. Indirect influences are disease and parasites. Excessive heat or cold increases the metabolic energy required to maintain the animal's body temperature, thus reducing the energy available for productivity. This energy imbalance is usually corrected by increased feed, which entails an additional cost to the farmer. The use of climatological data and analysis is useful in this case. Weather, Climate and Animal Performance provides many examples of using weather and climate information in the application of animal housing. It mentions that siting, external wind, temperature and humidity affect both entry conditions and patterns of internal air movement in environmentally controlled housing. As shown the importance of the ventilation rate in animal housing, and how it may be estimated by a psychrometric method simple enough to be applied on a farm scale. The siting of animal shelters is also important, particularly in relation to other shelters, as alternating positive and suction pressures may result from nearby obstructions.

10.3.2 Storage

For some crops, storage is as important as production and weather may affect the quality of the food product and modify the storage environment, causing loss of product and economic value.

Fruit and vegetables The Effect of Weather and Climate upon the Keeping Quality of Fruit states that the storage life of fruit is terminated by the onset of rotting caused by specific fungi, physiological diseases and senescent breakdown. The rate of development of these storage disorders is determined by storage conditions such as temperature and concentration of CO_2 and oxygen, and by pre-storage treatments. Agroclimatology of the Apple Crop states that the maturity of the apple fruit has a major impact on the quality of fruit during storage and that there have been many studies on modelling apple maturity using climatic data.

A WMO survey of operational models for agrometeorological services related to potato production states that monitoring of the storage environment and weather forecasts are used in many countries to determine the optimum conditions for potato tubers. In the former Czechoslovakia, humidity and temperature data (the mean number of hours exceeding the limits of 3°C, 7°C and 10°C during the winter months) were used for the design and construction of large potato storage facilities.

Grain: Agrometeorology can provide guidance for the construction of grain storage based on local climate modification and environmental control. Stored grain interacts with its environment, exchanging heat and moisture. The level of biological activity of grain and potentially damaging organisms must be minimized. For safe storage, grain must be kept cool and dry, requirements that are affected both by the characteristics of the building or structure for housing and by the external environment. Heat uptake from outside must be minimized while heat loss from storage must be maximized. The moisture exchange between the grain and the external environment should generally lead to a reduction in grain moisture content. When hot dry grain must be cooled to prevent insect infection, the resulting moisture increase must be kept to an acceptable level. The siting of storage facilities and their design and construction materials can all be influenced by meteorological factors. For example, in hot, dry regions with no refrigeration, there are advantages to storage facilities of high thermal capacity, with air space ventilated at night when the air is coolest. In warm, wet regions with small diurnal temperature ranges, good storage may be difficult to design. Meteorological factors become more important to the farmer when natural air-drying systems are used for grain. In the state of Ohio, United States, for example, best results for shelled corn occur when the climatic conditions provide temperatures in the range of −1°C to 10°C and relative humidity in the range of 60 to 70 per cent (Hansen et al., 1990). In most western Ohio counties, October and November provide high probabilities for good climatic conditions for natural air drying of soybeans and shelled corn. There is also an impact of weather conditions due to rainfall before storage that can be important. High temperature and rain that occur shortly before harvest are the most important direct weather effects on the quality of spring barley.

10.3.3 Averting Dangers to Production

These dangers may be the direct result of weather (for example, frost) or they may be indirect and carried by biological agents that are affected by the weather (such as pests and diseases that attack plants and animals).

Direct weather hazards

- ***Frost*** The occurrence of frost has been studied in detail by many agrometeorologists mainly because of its economic effect on high-value crops, and because some crops can be protected. Frost-risk maps and dates of first and last frost are simple but useful applications of climate data applied to agriculture. These maps are made at the macro- to mesoscale and are useful for specifying general planting dates for cereal crops and for the assessment of crop damage when combined with phenological data.

- **Sites** Assessment of potential sites for frost-sensitive crops, especially high-value crops such as tree fruit and coffee, is crucial since it will discourage growers from planting in frost areas. Topo climatology and local-scale agroclimatic zoning are important tools and methodologies in this regard. The Effect of Temperature on the Citrus Crop describes agrotopoclimatology as being concerned with the local differences in climate arising from topography, soil and vegetation within a uniform macroclimatic zone (this was defined earlier in this chapter as mesoclimate). They show some examples of using topoclimatological analysis to develop maps indicating the probability of frost occurrence over complex terrain. With the increase in availability and speed of personal computers in recent years, applications of this kind have increased. One example uses a spatial interpolation method to determine the spring frost hazard in the hilly areas of French vineyards based on digital elevation data and weather station temperatures.
- **Protection against frost damage** The two kinds of frosts that can affect crops and call for different protection techniques. Advection frost usually occurs during or after a change in airmass and is accompanied by strong winds (cold front). The number of protection techniques against this kind of frost is limited. Radiation frost occurs under the influence of a high-pressure system and typically the winds associated with this kind of frost are very light. They list the following frost protection methods: site selection, radiation interception, thermal insulation, air mixing, direct convective air heating, radiant heating, release of the heat of fusion and soil manipulation. Most of these methods are effective only against radiation frosts, but some can be applicable to both advection and radiation frosts.
- Techniques of Frost Prediction and Methods of Frost and Cold Protection describes many direct (or active) and indirect (or passive) frost protection methods, which are taken from mostly Russian and European sources. Direct methods of frost protection include: protective covers; smoke generation and artificial fogs; open-air heating of plants and areas; irrigation and sprinkling; and mixing air. Indirect methods include: biological methods such as hardening, seed treatment, selection of frost hardy strains, development of new frost-hardy varieties and regulation of bud development. Ecological methods such as control of mineral nutrition and crop site selection are other indirect approaches.
- **The Frost Protection** Fundamentals, Practice, and Economics lists many recommended methods of passive and active frost protection along with detailed practical overviews of each method. It describes recommended passive methods such as site selection, managing cold air drainage, plant selection, canopy trees, plant nutrition management, pest management, pruning, plant and soil covers, soil cultivation, irrigation, removal of cover crops, trunk wraps and painting, and bacterial control. Recommended active methods of frost protection include the use of heaters, wind machines, helicopters, various types of sprinklers, surface

irrigation, foam insulation, and some combination of these methods. It also provides a review of critical temperatures for annual, biennial and perennial crops, fruit and citrus trees, grapes, and other small fruits. A companion volume details several practical Excel software spreadsheets that help users to compute the probability that temperatures will fall below critical levels and the risk of frost damage specific to a crop , and to determine the economic risk of frost damage protection.

- **Hail** Hail can destroy high-value crops within a short time and many countries have therefore sought to reduce its frequency or to reduce the damage that it causes. Protection of Plants against Adverse Weather) contains a short summary of the protection methods that involve adding condensation nuclei to hail-forming clouds, by means of missiles or aircraft, with the object of producing small hailstones or soft hail. Large reductions in hail have been claimed by a number of groups. According to the WMO Commission for Atmospheric Sciences, the weight of scientific evidence to date is inconclusive, neither confirming nor refuting the efficacy of hail suppression activities.

Indirect weather hazards

Introduction to crop and animal pests/ diseases The application of meteorology to overcome the effects of pests and diseases on plants and animals involves a complete understanding of the complex life cycles of the pathogen and its host, as well as the environmental conditions that influence growth and development. Plant pathologists have developed a disease triangle with host (a susceptible animal or plant), environment (environmental conditions suitable for disease or pest establishment and development), and disease (the presence of the disease or pest) at the different points of the triangle, as depicted in Figure 1. These concepts help to describe the situation for virtually all known pests and diseases. All three sides of the triangle must exist for the pest/disease to be established and develop. If one of the sides is missing, then establishment of the pest/disease will not occur. Meteorological factors are very important for the growth and development of the host plant and animal species, for the pest/disease, and the airborne transport of the pest/disease. Lists temperatures, solar radiation, precipitation, leaf wetness and humidity, and wind as the major meteorological factors in relation to plant pathology. The duration of leaf surface wetness caused by dew, fog or rain is often a critical variable controlling the germination of disease spores. It must be computed from standard meteorological data or measured with leaf wetness sensors. Meteorology can be applied via observation of temperature, relative humidity and rainfall. More sophisticated applications include numerical weather prediction models for wind direction and speed with regard to the disease or pest on the desert locust and on foot-and-mouth disease). For the host or plant food source, temperature can be used for phenological development, rates of infection and disease/pest survival (extreme temperatures). Rainfall is important for host plant and disease/pest development. All of these aspects can then be modelled and used in operational applications for the agricultural decision-maker. One important use of weather and climate information is in the field of Integrated Pest Management (IPM). IPM strategies include avoiding the use chemicals unless there is economic damage to the crop. There have been numerous studies and models of the influence of temperature on plant and insect growth and development.

As provides many examples of agrometeorological aspects of operational crop protection, including protection against plant diseases, insect pests and weeds; meteorological data requirements; and application of weather forecasts. Definition of Agrometeorological Information Required for Vegetable Crops provides a brief overview of the meteorological factors relating to vegetable pests and diseases.

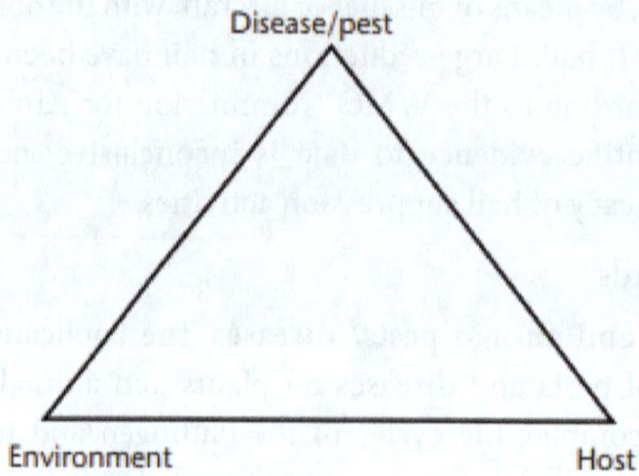

Figure 1. Disease/pest triangle

Crop pests: There are many applications of meteorology involving crop pests. Based on the concepts in the pest/disease triangle, applications focus on using temperatures to predict insect development and host plant development. Any additional weather parameters can then be added, depending on the nature of crop pest. During a desert locust outbreak or plague, the type and amount of meteorological information increases to include daily temperatures and forecasts of temperature and rainfall, as well as wind forecasts, which are especially important. Recently, more attention has been given to high-resolution regional numerical weather prediction models.

Other crop pests Most of the life of the beetle is spent beneath the ground as a grub. The appropriate environmental model therefore takes account of summer rainfall (as a substitute for soil moisture) and summer and winter soil temperatures. As describes the model used to consider the effect of agrometeorological conditions on the principal characteristics of Colorado beetle activity: duration of development, reproduction and death of individuals. The preliminary number of chemical treatments to be applied to potato is calculated on the basis of the forecast intensity of Colorado beetle reproduction, then it is corrected in the light of observed weather conditions. There have been many studies relating meteorological factors to important crop pests, such as the cotton leaf worm and pink bollworm, Colorado potato beetle, various pests of sugarcane and cassava mites.

Crop diseases The application of meteorology as an aid to the farmer in combating plant disease differs according to the mechanisms by which each pathogen is spread. The pathogen may be a year-round resident that increases and spreads whenever the weather is suitable for the pathogen and the host plant, which is the case with the fungal disease called potato blight, for example. In some areas a pathogen may not be capable of surviving the year, and may not reappear unless transported in sufficient quantity from a distant source, as in the case of black wheat rust, for example. In recent years the development of crop disease models has focused on crops with high economic value, such as fruit trees, vineyards and vegetables, since the models need meteorological observations in field settings that usually require the establishment of costly automatic weather stations.

- **Potato blight** There are two approaches available for the forecasting of late blight with a view to reducing agrochemical use compared to routine 7- and 10- day spraying:
    - Simple meteorological rules related to the life cycle of Phytophthora infestans that use rainfall, temperature and humidity over 12–48 h periods to predict spore production ("critical periods") and possibly subsequent periods when risk of infection is greatest. These methods can be used with either hourly observation data or synoptic weather maps;
    - Computer-based decision support systems (DSSs), usually utilizing the simple rules that rely on data from in-field automatic weather stations or available as digital files from (usually) Internet sources.

The rules differ only in detail and require some regional or site-specific calibration. It is indicated that ideal conditions for spore production are relative humidity above 95 per cent and temperature above 10°C at nighttime; free water must be available on the crop surface for serious infection to occur, so rainfall and prolonged high relative humidity are also required after spore production. Simple rule-based methods predict critical periods from late spring until late summer using the following general rules:

- Minimum temperature >10°C for a period of between 12 to 48 hours;
- Relative humidity >90 per cent for the same time period;
- Rainfall in the period following (4 hours to 10 days later).

Evidence of disease is expected between 7 and 21 days after a critical period has been predicted and a suitable crop protection strategy can be put in place. If possible, prediction should be based on hourly observation of temperature and relative humidity. Synoptic maps can be used (particularly where observations are sparse) to make predictions based on the likelihood that current and forecast weather systems will create suitable conditions for a critical period to occur, such as the passage of warm, moist tropical air giving rise to high humidity and temperature, and slow-moving depressions giving rise to overcast, humid, rainy conditions. The use of DSSs typically requires site specific automatic weather station data, but interfaces with alternative data sources are also possible. The systems automate the use of the simple rule-based methods and usually provide spray strategy recommendations as well.

- **Wheat diseases** It provides an overview of the various wheat rusts that occur around the world and the meteorological factors that contribute to the transport of spores and disease outbreaks for various types of wheat rusts. Also provides a survey of many crops' disease models, including several for wheat. It describes EPIPRE as a system devised to support decision-making in pest and disease control in winter wheat with the aim of reducing pesticide use. The system integrates six fungal diseases and three aphid pests of wheat. Spirouil–Epure was developed in France and has been used for many years to support extension services for brown wheat rust. The model uses meteorological data, along with some agronomic and phenological data, and provides advice on the dates for first fungicide application within microregions and well-defined crop zones.
- **Apple scab** Apple scab is caused by the *Venturia inaequalis* fungus and is an economically important disease for apple producers. The Influence of Weather Conditions on the Occurrence of Apple Scab provided one of first overviews of the disease, investigations in various countries and descriptions of the early warning systems. More recently, the apple scab model ASCHORF was developed in Germany. This model can be used to provide practical recommendations to plant protection services and apple growers). The modelled infection risk is dependent on temperature and leaf wetness duration. Leaf wetness duration is calculated but not measured and is based on energy balance principles. The model uses a sliding 10-day time series by inputting data for the previous four days from the standard meteorological network and then inputting grid point data from numerical weather prediction models.
- **Downy mildew** Downy mildew *(Plasmopara viticola)* is one of the most important fungal diseases for wine grapes (*Vitis vinifera*) and can lead to considerable losses in grape yield and quality. Friesland et al. (2005) developed the PERO model to calculate the start of infection of the grapevine disease Peronospora, which is determined by temperature and leaf wetness. The PERO model is based on laboratory and field experiments and the inputs are hourly air temperature, relative humidity, calculated leaf wetness, daily extreme temperatures and daily rainfall. The model outputs are infection dates and oil spot balances (lesions), which arPLASMO (Plasmopora simulation model) was developed to simulate the biological cycle and the disease leaf area of grapevine downy mildew, allowing

for the best timing for fungicide treatments, Data inputs are hourly temperature, relative humidity, rainfall and leaf wetness. The results are expressed in percentage of leaf area covered by oil spot lesion. The PLASMO model has been developed into a computer program for distribution and is also available on the Internet for greater access. Weather data can be uploaded to the model Website for running of the model used for agrometeorological advice.

Table 10.1 Examples of DSS-type systems for prediction and management of blight outbreaks (after Bouma and Hansen, 1999)

Model	Country and e-mail	Original development year	Main target users	Input	Output
Televis	Norway Arne.Hermansen@ planteforsh.no	1957	Farmers, advisers	Weather data	Epidemiological data
Guntz-Divoux	France fredec.nord.pas-de-Calais@wanadoo.fr	1963	Adviers, extension service	Weather data	Advice line
Slimphyt	Germany	1982	Plant protection service, extension officers	Weather data field data	Field data Advice
ProPhy	Netherlands	1988	Farmers, advisers, extension officers	Weather data field data other data	Weather overviews Field data Advice
Plant-Plus	Netherlands	1990	Farmers, advisers, suppliers, processors	Microclimate Crop + product information	Fungicide protection periods
I.P.I	Spain	1990	Farmers, advisers	Weather data	1st spray timing Epidemiological data
NegFry	Denmark JensG. Hansen@agrisci.dk	1992	Farmers, advisers	Wather data Field data	1st spray timing Fungicide applications
PhytoPRE + 2000	Switzerland Hansrudolf.forrer@ fal.admin.ch	1995	Farmers, advisers, plant protection service	Weather data Field data\Other	Regional data Field
Guntz-Divoux	Belgium pcg@ping.be	1996	Advisers, extension service	Weather	Advice line

10.3.4 Other Applications

Norway has developed a Web-based site-specific warning system called VIPS that calculates warnings for several pests and diseases in selected fruits, vegetables and cereals. The warnings are linked to over 70 weather stations and colour-coded warnings ranging from danger (red) to no danger (green) are given for each county for the previous five days and are forecast for the upcoming five days. VIPS incorporates previous work done in Norway on pests and diseases such as NORPRE. NORPRE is a cereal disease and pest control system that uses daily weather data as input to a number of different submodels. The system uses field observations of pest disease occurrence from farmers to validate the models and adjust threshold values. The system includes the following pest models: cabbage moth, turnip moth, carrot fly and codling fly.

Table 10.2 Examples of "critical period" prediction methods indication the types of modification made for various regions where potatoes are grown (not an exhuastive list)

Method	Temperature	Humidity	Ohter	Rainfall	Reference
Dutch rules 1926 Europe	>10oC min at night	4 h below dewpoint at night	8/10ths following day	Followed by >0.1 mm rain	Van Everdingen (1926)
Beaumont periods 1947 Uniterd Kingdom	>10°C min for a minimum 48 h period	75% during the 48 h period			Beaumont (1947)
Irish Rules 1953 Ireland	>10oC min for a 12 h period use temeratur bands and	>90% for the 12 h	Free moisture on leaves for 2 h after the 12 h period or the rainfall criterion	Between the 7th and 15th h around the end of the 12th h of the 12 h period	Keane (1982)
Hyre rules 1954 North-eastern United states	5-day average <25.5oC excluding days with minimum <7.2oC		Looks for 10 consecutive risk days	Total rain in 10-day period >30 mm	Hyre (1954)
Smith periods 1956 United Kingdom and Ireland	>10°C min for 2 x 24 h periods	>90% for 11 h in each of the 2 periods			Smith (1956)
Negative prognosis 1966 Germany	multiplies the hours in each band by a weighting factor	At low temperature only use hrs when >90% or a rainfall limit for 4 h blocks. At higher temperature use a 10 h block	Can subtract from risk when RH is <70%	At low temperature only use h Ullrich and when >0.1 mm rainfall or RH limit for 4 h blocks. At higher temperature use a 10 h block	Schrodter (1966)
Young rules 1978 South Africa	>10°C and <24°C >10°C and <24°C for 2 x 24 h periods	>70% at 2 p.m. during each of the 24 h periods			van Rij and du Preez (2004)
Forsund rules 1983 Norway	Maximum 17°C-24°C Minimum >10°C for 2 x 24 h periods	>75% at noon during each 24 h period		>0.1 mm during each 24 h period	Forsund (1983)
Winstel rules 1993 Belgium	Phase 1: Average daily >10°C and <23°C for 10 h and then 10 h >10°C Phase 2: Maximum daily >23°C and <30°C for	Phase 1: >90%			Winstel (1993)
Washington State rules 1996 North-western United States	Rainfall indicators calculated for periods when minimum >5°C during April and May and then July and August		Calculates probability of a year being an "outbreak" year using discriminant functions and binary logis¬tic regression	Days with >0.25 mm and temperature >5°C	Johnson et al. (1996)

There have been many surveys of crop disease models undertaken over the years. A WMO report (WMO, 2000d) contains a survey focused on fungal pathogens and lists 58 crops and 133 pathogens.

10.3.5 Animal Pests/Diseases

The approach of the meteorologist to problems of animal disease is basically the same as for plant disease. Weather and Animal Diseases provides a good overview of practical links between weather and animal disease that may be windborne, parasitic, fungal or the result of environmental or nutritional stress. More recent reviews detail internal animal parasites and cold and hot weather stress. Weather, Climate and Animal Performance states that there are two lines of enquiry in using climate information as a measure of disease incidence. The first uses climatic factors to develop climatic indices known to influence the development of the animal parasite during its life cycle outside the animal. The second uses biological development rates, calculated from the study of parasites under laboratory conditions in constant temperature chambers, to determine the influence of temperature variation on parasite development in actual field conditions. Besides the animal diseases listed below for which meteorological applications have been developed, this publication also provides information on nematodiriasis and parasitic gastro-enteritis.

- **Foot-and-mouth disease** The 1968/1969 foot-and-mouth disease epidemic in the United Kingdom led to research indicating that meteorological factors are major contributors to the spread of the virus during a foot-and-mouth outbreak (Smith and Hugh-Jones, 1969; Wright, 1969). The virus is spread when it is exhaled by animals as the nuclei of water droplets, which are then dispersed by winds. The most important meteorological factors are wind, humidity and rainfall; the synoptic situation will determine to some degree the distance over which the virus can spread. Wind speed and direction will determine the pattern of dispersion of the virus. Stable atmospheric conditions favour dispersion over large distances because vertical distribution is minimized, while high winds and turbulence usually reduce the transport range. Humidity will determine the duration that the virus remains protected by its water droplet. Maximum infectivity is associated with relative humidity above 60 per cent (Murphy et al., 2004). Rainfall will influence when the virus is deposited from the atmosphere. In rainy conditions the virus will be deposited on herbage within short distances of the source, rather than moving to infect other animals.
- The more recent epidemic in the United Kingdom in 2001 resulted in the development, utilization and testing of models for predicting the spread of virus based on: the predicted viral load at a source location; the predicted spread due to surface weather conditions and latitude and topography. It is noted that older models depend on synoptic observations and thus suffer error due to potential remoteness of observation stations from outbreak sources. NAME and DERMA

are long-range models driven by numerical weather prediction (NWP) output, and produced similar results despite being driven by different NWP models. The local-scale models, driven by nearby observation data, were also used to analyse local infection. It was concluded that 24-hour average virus concentrations do not adequately represent infection risk and that short-term high concentration levels are needed to account for the pattern of infection that was observed. For local-scale prediction, the most important observation/NWP output is 10 m wind speed, estimates of three-dimensional dispersion, the relative humidity and the chance of rainfall occurring. For regional-scale modelling, 1- to 3-hour NWP output is preferred and should include wind speed, wind direction, relative humidity, cloud cover and precipitation.

- **Facial eczema of sheep** A warning system for this fungal disease was devised in New Zealand. Even before a definite link was established between the disease and fungus present in grass, soil temperatures and rainfall were used for warning. High humidities and ambient temperatures in the 21°C–27°C range are favourable for the spores, and the discovery of the fungus reinforced the empirical approach. Spore traps and counts are now being used to confirm the meteorological evidence.
- **Fascioliasis in sheep** Fascioliasis (commonly called liver fluke disease) is a parasitic disease that affects sheep and is caused by *Fasciola hepatica*. The complicated life cycle consists of the passing of fluke eggs in dung by infected sheep, and these eggs then hatch into free swimming larva in the open pasture and infect the fluke's intermediate host, a snail, *Lymnea truncculata*. This is the most sensitive stage of the life cycle. The larvae will die if they cannot enter a snail within 24 hours. Also describes a wetness index (Mt) based on monthly rainfall, potential transpiration and the number of rain days, which was developed in England and Wales. Data for the index are accumulated over a season, and based on comparison with historical disease statistics, thresholds for treatments can be established.

10.4 OTHER APPLICATIONS

There are many other applications of meteorology for agriculture besides those already mentioned. They are covered in other chapters in detail because of their importance. One important group relates to the physiology and growth of plants, from germination to final yield. These are affected somewhat by the applications already dealt with, for example, irrigation, shelter, cover and disease. Other applications can be cited, including the use of degree-days or other indices to determine the phenological stages of crops, such as flowering, reproduction and maturity. These stages are very important for pest/disease management. Typically, growing degree-day or heat-unit calculations are made by subtracting a threshold temperature from the average daily temperature and then accumulating these units over time to model plant and insect development. The simplest form on a daily basis is:

$$\text{Degree-day} = [(T_{max} + T_{min})/2] - T_{base}$$

It provides a good overview of modelling crop phenology for many crops. Weather Based Mathematical Models for Estimating Development and Ripening of Crops describes these calculations as temperature-remainder models (TRIM) and lists many methodologies to calculate crop development. See Chapters 6 and 10 for a more detailed review of these concepts. Another group of applications concerns field operations. Since cultivation, drilling, spraying and harvesting are all highly weather-dependent, the meteorologist can give considerable help in assessing the probability of weather suitable for these operations, which may greatly affect the requirements for labour and machinery. Also showed that lodging results in a significant crop yield drop (by as much as 20–30 per cent) and degradation of grain and straw quality, and also creates problems for harvesting. Operation of harvesting units with lodged crops is extremely difficult and their efficiency decreases by 25–50 per cent, which results in longer harvesting times. But even under optimum harvesting conditions, the grain loss is in the range of 10–25 per cent. Depending on the lodging conditions, biological losses are between 5 and 40 per cent on average. In all applications meant for the farmer, it is of the utmost importance that the meteorologist work closely with the specific experts on individual problems.

11

WEATHER AND AGRICULTURE RELATIONS

11.1 AGRICULTURE AND WEATHER

- Weather includes many factors such as sunshine, temperature, precipitation, evaporation, transpiration etc. Agriculture and weather relations explain as following:

  - ***Solar Radiation:*** It includes light intensity, light quality and duration of sunlight. Out of total radiation received, only 50% in photsynthetically active radiation (PAR) which lies in 400-700nm range. Rest is UV or IR. Now there is an exponential relation between amount of light intercepted by canopy and Leaf Area Index (Leaf Area/ground area). The sum of these values for individual days is directly proportional to crop yield.
  - ***Temperature:*** A term growing degree days is used which is given for a crop from time of flowering to harvest date. Every crop has a temperature range below or above which GDD for that particular day is zero. GDD is the sum of difference of daily temperature and base temperature. So the harvest date is predicted depending on the when the GDD is achieved.
  - ***Precipitation,Evaporation and Transpiration:*** There is a relation to calculate the length of growing period which includes Evapotranspiration. You need to know amount of precipitation and crop water requirement. Also transpiration causes cooling so maintains temperature. This prevents crop from damaging.

- These factors are responsible for proper crop growth and development. If anyone of the factors is affected, the crop will be affected accordingly as in case of heavy rains, high temperatures etc.

11.2 ROLE OF WEATHER AND WEATHER FORECASTING IN AGRICULTURE

Weather has a significant impact on the agriculture industry. Indeed, crops need the basics of moisture, warmth, and sun to thrive. But what's less obvious is how the details of weather information can drive a grower's business decisions, helping them to plan efficiently, minimize costs and maximize yields—and profits—as a result. Four primary areas of farming that are fundamentally affected impacted by weather:

11.2.1 Crop Growth/ Irrigation

Crop growth, or crop yield, requires appropriate amounts of moisture, light, and temperature. Detailed and accurate historical, real-time and forecast weather information can help farmers better understand and track the growth status/stage to make informed decisions. Having access to this data can guide farmers in making significant and potentially costly decisions, such as whether, when and how much to irrigate.

11.2.2 Fertilizer Timing and Delivery

- One of the many decisions that farmers have to make is determining the proper

Agriculture is highly dependent on prevailing weather conditions. Significant fluctuations and increased frequency of extreme weather events such as floods, droughts, heat and cold waves, hailstorms, strong winds, cyclones etc. have made agriculture more vulnerable to climatic risks. Global warming and climate changes are expected to increase in future. All these changes are likely to have severe implications on agricultural production especially in the tropical and sub-tropical regions including India. Increase in temperature is likely to have adverse effect on crop productivity due to shortened crop duration and increased respiratory losses. Erratic precipitation patterns are expected to enhance year-to-year yield fluctuations because of increased frequency and intensity of droughts and floods. Microclimate, referring to the climatic elements in the immediate vicinity of the plants, is very important as it regulates and determines the plant physiological reactions and energy exchange processes between the plant and its surrounding atmosphere. Lack of optimal climatic elements leads to disturbances in these processes and results in undesired decrease in crop productivity. Microclimatic modifications help in modifying the adverse conditions prevailing in the immediate vicinity of the plants making it favourable for the better crop growth, development and yield. Artificial control of field microclimate to maintain the optimum conditions for better plant growth and crop production can be achieved by making field level adjustments. Short term adjustments at the farm level such as appropriate sowing time, row spacing and orientation, changes in crop rotation and crop cultivars, changes in soil cultivation and tillage practices, planting method, mulch application, use of shelterbelts / wind breaks and intercropping etc. result in the maintenance of favourable crop microclimate by moderating temperature extremes, conserving soil moisture and increasing radiation interception. Under the present scenario of global warming and increased occurrence of extreme weather events, adoption of such microclimatic modifications in crop production is the need of the hour to manage extreme weather risks and improve crop productivity to attain food security and sustainability of natural resources under changing climatic conditions.

time to apply fertilizer, as well as the application rate and fertilizer form to use. A misapplied application caused by weather can wipe away the entire field's profits.

- Weather forecasts can be used to ensure that fertilizer is applied in the right conditions—when it's dry enough so that it doesn't wash away (which would create a waste of resources and money) but moist enough so that it gets worked into the soil.

11.2.3 Pest and Disease Control

- Certain weather conditions encourage the development and growth of pests and diseases, which can destroy crops. Forecast guidance incorporated into pest and disease modeling can help determine whether—and when—it's appropriate to apply pest or disease controls.
- Wind forecasts also play a role in this decision, as crop dusters, aircraft that spray fungicidal or insecticidal chemicals on plants from above, must be utilized when wind conditions are not apt to cause sprayed chemicals to miss their targets.

11.2.4 Field Workability

Field workability refers to the availability of days that are suitable for fieldwork. It's primarily dependent upon soil moisture and soil temperature. Accurate field-level weather information can help farmers assess the workability of their fields and become more efficient in their day-to-day operations.

The cost/benefit equation for having access to reliable weather forecast information is not always easy to quantify, but it's a decision that's easy for most large growers and producers to make. Throughout many months, farmers make small but frequent decisions about their crops, and the cumulative effect of the financial implications of those decisions can be significant.

Irrigation planning is a good example. If a farmer relies on a forecast for precipitation that turns out to be accurate, he saves the cost of unnecessary irrigation. And by having a good idea of the expected amount of rain over a period and irrigating just enough to allow crops to thrive, he will maximize yield.

11.2.5 Role of Weather Forecasting in Precision Agriculture

- Weather information is playing an increasingly instrumental role in the evolving field of precision agriculture, a farming practice that emphasizes accuracy and control when it comes to the growing of crops.
- An essential aspect of this approach is the use of information technology, which includes weather prediction and other items, such as satellite and aerial imagery, GPS guidance, sensors, drones, variable rate fertilizer application, and crop health indicators.

- The ultimate goal of precision agriculture is to maximize growth efficiency at the individual seed and plant level. As a "big data" weather provider, WeatherOps is excited to provide hyper-local weather data and knowledge that's being used to help make this goal a reality.

11.3 MODIFICATIONS OF CROP MICROCLIMATE

Microclimate

- A microclimate is a local set of atmospheric conditions that differ from those in the surrounding areas, often with a slight difference but sometimes with a substantial one. The term may refer to areas as small as a few square meters or square feet or as large as many square kilometers or square miles. Because climate is statistical, which implies spatial and temporal variation of the mean values of the describing parameters, within a region there can occur and persist over time sets of statistically distinct conditions, that is, microclimates.
- Microclimates can be found in most places. Microclimates exist, for example, near bodies of water which may cool the local atmosphere, or in heavy urban areas where brick, concrete, and asphalt absorb the sun's energy, heat up, and re-radiate that heat to the ambient air: the resulting urban heat island is a kind of microclimate.
- Another contributing factor of microclimate is the slope or aspect of an area. South-facing slopes in the Northern Hemisphere and north-facing slopes in the Southern Hemisphere are exposed to more direct sunlight than opposite slopes and are therefore warmer for longer periods of time, giving the slope a warmer microclimate than the areas around the slope. The lowest area of a glen may sometimes frost sooner or harder than a nearby spot uphill, because cold air sinks, a drying breeze may not reach the lowest bottom, and humidity lingers and precipitates, then freezes.
- Increase in the frequency of extreme weather events due to global warming triggered climate changes have resulted in manifold increase invulnerability and climatic risks in agriculture. Inter- and intra-seasonal weather variability and extreme weather events like droughts, floods, heat and cold waves, strong winds, hailstorms, cyclones etc. have increased over the recent decades .Heat waves or extreme temperature events are projected to become more intense, frequent and last longer than what is being currently observed . Although climate change also occurs due to natural processes such as variations in solar output, earth's orbital changes and volcanic eruptions etc., but at present, climate change is happening mainly due to the human activities. Total anthropogenic green house gas emission shave been observed highest in the human history during the period 2000-2010 (IPCC, 2014). Major processes which contribute towards climate change are

burning of fossil fuels, industrial processes, deforestation and agriculture. Among various green house gases, carbon dioxide contributes by 76%, methane by 16%, nitrous oxides by 6% and chlorofluorocarbons by 2%. Agriculture contributes 28% of the Indian greenhouse gas emissions primarily through methane emission from paddy fields, enteric fermentation in ruminant animals and nitrous oxides from application of manures and fertilizers to the soil .

- As a consequence of these processes, global temperature has increased by 0.85°C during last100 years. Average precipitation has increased over different regions of earth. These changes have resulted in melting of polar ice and glaciers leading to rise in sea levels and floods over various regions (IPCC, 2014). Global warming and climatic changes are expected to increase in future. Global circulation models have predicted rise in global average temperature by about 2°Cby 2100. Although total precipitation is also predicted to increase during kharif season, but trends may vary at local level. Similarly, atmospheric CO2 concentration is projected to increase from 478 to 1100 ppm by the end of the21st century (IPCC, 2007). Under such conditions, snow cover is expected to contract and frequency and intensity of extreme weather events like heat and cold waves, intense rainfall events etc. are likely to increase. There are evidences of increased heavy precipitation and decreased light precipitation in widespread parts of the globe due to global warming . Due to highly erratic rainfall, there is an increased risk of drought as a result of increased prolonged dry spells, total dry days and decreased light precipitation days over India as a consequence of global warming .
- All these changes are likely to have severe implications on agricultural production especially in the tropical and sub-tropical regions including India. It is reported that warmer temperatures and more extreme temperature events will significantly impact plant productivity. Frost cause sterility and abortion of formed grains, while excessive heat cause reduction in grain number and reduced duration of the grain filling period (Barlow et al., 2015).Under these scenarios, crop production in India is likely to decrease by 10-40% despite the beneficial effects of higher CO_2 on crop growth. As rabi season crops are adapted to cooler conditions, thus greater losses are expected in rabi crops. Every 1oC increase in temperature can reduce wheat production by 4-5million tonnes in India, however, timely planting can reduce these losses to 1-2 million tones. Similarly uncertainty in rainfall leading to droughts and floods is likely to enhance year-to-year yield fluctuations. However, reduced frequency of frost under warming scenarios may prove beneficial for frost sensitive crops like potato, peas etc.
- Agriculture is highly dependent on climatic conditions, but significant fluctuations and increased frequency of extreme weather events have made it more vulnerable by increasing climatic risks in crop production. Under such conditions, there is a

dire need to manage climate change impacts on crop production to ensure food security for burgeoning population along with sustainability of natural resources. Since rise in temperatures are likely to reduce crop yield, thus it is imperative that suitable adaptation strategies have to be developed to minimize the adverse impacts. It is reported that location specific and integrated farming system based technological management options reduce the climatic risks and better utilization of available natural resources result in higher agricultural productivity and thereby enhance food and livelihood security of small and marginal farmers. Microclimatic modifications can be an effective adaptation strategy in this direction. Microclimate refers to climatic conditions in a smaller area i.e. a few meters above or below the earth surface or within the crop canopy. It is the local climatic condition near the ground or area around the plants (up to about 2m height) resulting from the general climatic conditions. It also refers to the climate of small regions influenced by the effects of the relief, topography and the lower surface features, which create disparity between soil and air temperature, humidity and wind speed. Growth/development of plants mainly depends on their genetic constitution and environmental conditions. Climatic elements in the immediate vicinity of the plants are very important as they regulate and determine the physiological reactions. Lack of optimal climatic elements leads to disturbance in plant physio-logical and energy exchange processes resulting in undesired decrease in crop productivity. Under such conditions, there is a dire need to adopt microclimatic modifications to manage climate change impacts on crop production to ensure food security for burgeoning population along with sustainability of natural resources.

- Microclimate of the crop varies from top to bottom of the canopy. All the crop management practices namely sowing time, planting method, row spacing, intercropping, tillage practices, mulch application, shelter belts and irrigation management etc. affect the microclimate due to their effect on canopy temperature, wind speed, soil moisture, light interception and rate of waterloos etc. In other words, these microclimatic modifications affect the rate of exchange processes within the canopy as well as between the canopy and the surrounding air. Thus, by making some alterations / adjustments in crop management, we can modify the crop microclimate without any significant financial burden. Microclimate modification includes an artificially induced changes in the composition ,behaviour or dynamics of the atmosphere near the ground so as to improve the environment in which the crops are grown. By making such alterations, the microclimate can be made more favourable for better growth and yield of the crops.

11.4 PROCESSES OF MICROCLIMATIC MODIFICATIONS

- Microclimatic modification is any process in which all the adverse or stress conditions prevailing in the immediate vicinity of the plants are modified for the better crop growth, development and yield. Artificial control of plant environment to maintain the optimum conditions for better plant growth and crop production is future trend in agro meteorological research. Short-term adjustments at the farm level involve production techniques, such as changes in crop rotation and crop cultivars, changes in soil cultivation and tillage practices, a shift of sowing dates, adapted fertilization and crop protection measures. Long-term adaptations, on the contrary, include major structural changes of farm production systems and need careful agro-economic planning and realization at societal level; these adaptations also involve a set of sectors and stakeholders, such as policy, research, water and land planning. Some examples of long-term adjustments are changes in land use and landscape structure, breeding and biotechnology applications, crop substitution and changes in the farm production type. A change in the landscape structure, such as the introduction of windbreaks or hedges, influence the microclimate of crops in neighbouring fields, mainly by slowing down the wind speed. Further effects include increased dew formation and leaf-wetness duration and a reduction of dry air advection, evapotranspiration, unproductive waterloos and wind erosion .
- Major processes of microclimatic modifications include controlling of heat load, water balance, wind speed and modification of temperature and solar radiation. Heat load on crop canopy can be increased (during winter) called heat trapping, or decreased (during summer) called heat evasion. Heat evasion can be achieved by shading the crop, irrigation management or by using anti-transparent, whereas heat trapping can be achieved by adopting appropriate row direction or plant density . Controlling the water balance involves increasing the amount of water stored in the root zone (by strip cropping, contour cropping, ploughing, terracing or bunding), increasing infilteration (by increasing tillage or increasing row spacing), reducing soil evaporation (by using mulches and crop residues, plastic covers or thick cultivation crops),modifying transpiration (by using anti-transpirants, substances forming film on leaf surface, stomata closing materials and reflectants).Whereas, winds can be controlled by applying wind breaks or shelter belts. These protections reduce wind speed on the leeward side and save the plants from freezing injury and mechanical damage (Fig. 1).

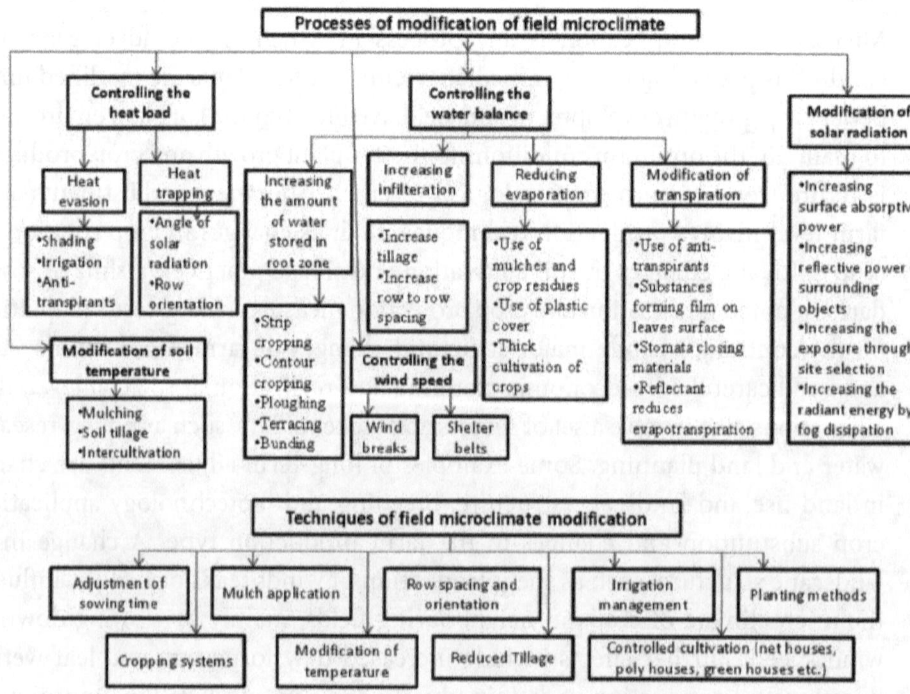

Fig. 1. Processes and techniques of field microclimate modification

Modification of solar radiation can be accomplished by increasing the surface absorptive power, reflective power of the surrounding objects, exposure through site selection, increasing the radiant energy by fog dissipation and by adopting appropriate row direction. Similarly, soil temperature can be modified by mulch application and soil tillage etc. Mulch helps in reducing the soil temperature in summer and increasing during winter season. Similarly soil tillage by harrowing or inter-cultivation also helps in modifying the soil thermal regime. Similarly, protected cultivation under controlled environment like green houses(for temperature, humidity and CO_2 control), polyhouses (naturally ventilated or with controlled environment) and shade houses (for crop production in warm climate) provide other means for raising crops under controlled microclimatic conditions.

11.5 MAJOR TECHNIQUES OF FIELD MICROCLIMATE MODIFICATIONS

- Different crops have specific climatic requirements at different growth stages. Thus, raising crops under favourable microclimate as per requirements of the crops, needs different management techniques so that sensitive growth stages of the crops can coincide with the favourable micro-environment.
- Various agronomic manipulations and other microclimate modification options viz. alteration in sowing time, mulch application, row orientation and spacing,

irrigation management, planting methods, tillage practices, appropriate cropping systems and installation of the net houses/screen houses etc. can be used to make the crop microclimate optimum for highest growth rate and productivity.

11.5.1 Alteration in Sowing Time

- Sowing time is an important production component that can be manipulated to counter the adverse effects of environmental stress. This is accomplished through shifting of sowing time so that any stress caused by environment is avoided during the critical stages of crop growth. Matching the phenology of the crop to the duration of favourable environmental conditions by selecting the most appropriate sowing time to avoid the periods of stress is crucial for obtaining maximum yields under changing climate . Favorable microclimate can be created by modifying the date of sowing, which takes care of the deleterious effect of high temperature during reproductive growth period. Temperature is one of the most important elements of climate which determines directly the potential productivity level particularly for winter crops. Light response i.e. photoperiodism, which not only control the temperature factor but also the vegetative growth as well as flowering of the plants, is important weather element for wheat crop to assess the thermal response and its requirement at different phenophases to harvest the potential yield . For getting higher yield, sowing time of crops needs to be adjusted with suitable agro climatic environment .
- Terminal heat stress intensity is severe under late sown conditions in Punjab causing reduction in the duration of later growth phases and hence grain yield . Substantial increase in grain yield of wheat can be achieved by sowing the crop at the optimum time. Heat shock at the end of tillering strongly reduces the rate of leaf photosynthesis, while during grain filling it decreases both rate of photosynthesis and grain growth . The grain yield of cereals has been reported to increase with increase in allocation to the reproductive organs with little increase in biological yield (Donald and Hamblin,1976). It would be, therefore, appropriate that plants function in such a manner that maximum amount of dry matter goes to the spikes for increasing weight of grain during post anthesis period, leading to higher grain yield. Partitioning of dry matter at anthesis indicates the maximum contribution from stem, followed by roots, leaves and spikes, whereas at physiological maturity, the spikes contribute maximum, followed by roots, stem and leaves. Date of sowing influences the yield considerably and delays in sowing subject the crop to mature early due to rise in temperature resulting in decreasing the number and size of grains. Wheat being a thermo sensitive crop, choice of suitable cultivar for different seeding time further gets prime importance.

11.5.2 Mulch Application

- Mulch is a layer of material applied at the soil surface, which leads to conserve soil moisture, moderate soil thermal regime, reduce weed growth and improve fertility and soil health. It is observed significant increase in grain and straw yield of wheat when the crop residues were retained in the field than when they were removed. It is observed increase in water use efficiency of maize with increase in mulch amount from 2 to 6 Mgha^{-1}Mulch obstructs the solar radiation reaching the soil surface. They increase the soil temperature during winter and increase during summer season. White- and light-coloured materials which increase the reflectivity, are used to decrease soil temperature during summer season, whereas, dark-coloured materials and black plastic mulches are used to increase soil temperature during winter season. Straw mulches can be used during both the seasons as they increase soil temperature during winter season and decrease during summer season. Due to obstruction of the exchange of water vapours, mulches decrease the rate of evaporation from the soil and conserve soil moisture for use by the plants.
- Hou et al. (2010) examined the effect of duration of plastic mulching on soil temperature, evapotranspiration, growth, yield and water use efficiency of potato under drip irrigation in an arid region of Northwest China. Daily mean soil temperature under the plastic mulch was 2-9 °C higher than under non-mulching conditions. The mulch effect on soil temperature was the greatest during the early growth and reduced as the plant canopy increased. Mulch reduced irrigation water required and evapotranspiration, however, extending mulch duration beyond 60 days had little effect on evapotranspiration. Both tuber yield and water use efficiency benefitted from early plastic mulching. Mulch cover for 60 days was found favourable for potato production compared to potatoes grown without mulch.

11.5.3 Row Spacing and Orientation

- Row spacing has a great impact on canopy temperature and radiation interception by the crop. If row spacing is more, radiation interception by the crop decreases, more radiation falls on the soil surface and increases the soil temperature. On the other hand, if row spacing is less, radiation interception by the crop increased and its transmission towards soil surface decreases, which decreases the soil temperature. David et al. (2002) reported that plant architecture changes affect the canopy light environment, yield and yield components of narrow-row cotton. Alterations of canopy architecture that allow more light penetration into the lower depths of the canopy may be a way to increase the yield of cotton through greater boll production.

- Appropriate row spacing and orientation help in trapping of radiation by the canopy, thus leading to higher radiation interception and crop productivity. Sandler et al. (2014) observed higher light interception with 19 cm row spacing in wheat as compared to that with 38 cm. Pathan et al. (2006) observed significantly higher light interception in wheat and barley along east-west row orientation as compared to north-south. Jha et al. (2012) reported significantly higher season accumulated IPAR in east-west as compared to north-south oriented plots in mustard. It is observed that 11% higher grain of wheat in the north-south sowing as compared to the east-west sowing due to its significant effect on crop microclimate. Significantly higher light intensity was observed in north-south direction, whereas canopy temperature was significantly higher in east-west direction.

11.5.4 Irrigation Management

- By applying irrigation, soil moisture availability increases, which increases the evapotranspiration and decreases the soil as well as canopy temperature. Under very hot and dry conditions, light irrigation is a common practice to save crops from high temperature stress. Alberto et al. (2009) compared the CO2 and heat fluxes in flooded and non-flooded rice fields in the Philippines to monitor the environment impact, in terms of carbon budget and heat exchange, of shifting from lowland rice production to aerobic rice cultivation as an alternative to maintain crop productivity underwater scarcity.
- The aerobic rice fields had higher sensible heat flux (H) and lower latent heat flux(LE) compared to flooded fields. On seasonal average, aerobic rice fields had 48% more sensible heat flux while flooded rice fields had20% more latent heat flux. Consequently, the aerobic rice fields had higher Bowen ratio (0.25) than flooded fields (0.14), indicating that a larger proportion of the available net radiation was used for sensible heat transfer or for warming the surrounding air. The total carbon budget integrated over the cropping period showed that the net ecosystem exchange (NEE) in flooded rice fields was about three times higher than in aerobic fields, while gross primary production (GPP) and ecosystem respiration (Re) were 1.5 and 1.2 times higher, respectively. The high GPP of flooded rice ecosystem was evident because the photosynthetic capacity of lowland rice is naturally large. The respiration (Re) of flooded rice fields was also relatively high because it was enhanced by the high photosynthetic activities of lowland rice as manifested by larger above ground plant biomass. The ratio of Re/GPP in flooded fields was 0.67 while it was 0.83 for aerobic rice fields.
- Appropriate amount and time of irrigation results in managing heat and water stress and in improving yield and water use efficiency. It is reported that the radiation use efficiency was more sensitive to water stress at mid-season (MS) than late season (LS) stage in legumes. The reduction in radiation use efficiency

during MS was 39, 30 and 29%, whereas during LS it was18, 19 and 17% for beans, chickpea and cowpea, respectively. Also observed that higher PAR capture ratio and radiation use efficiency with increase in irrigation application in maize.

- It also reported that increase in irrigations from one to four along with mulch application enhanced intercepted photo-synthetically active radiation in potato and increase in yield and heat use efficiency of wheat with increase in irrigation application.
- As indicated that canopy temperature depression increased with increase in irrigation application. canopy-air-temperature difference values were positive for most of the growing period for the rainfed and negative for well-watered treatment for both the years. The rainfed treatments under both flat and bed planting methods experienced the highest Bowen ratio, whereas the well-watered treatments experienced the lowest Bowen ratio, indicating that much of the available energy was partitioned towards sensible heat under rainfed treatment and towards latent heat for well-watered treatment. It also observed that highest water uses efficiency of wheat sown during mid-November with three post-sowing irrigations.

11.5.6 Planting Method

- Method of planting also affects the crop microclimate due to its effect on soil properties. It observed that beds enhanced10% yield of wheat due to improved soil physical condition as compared to flat sowing. It is noted that water saving of 26-42% in beds as compared to flat sowing in various crops. Bed sowing produced significantly taller plants with greater biomass and grain yields probably due to better nutrient availability, good soil conditions and weed control in beds.
- It is reported that furrow seeding improves the photosynthesis and dry matter production in cotton and observed higher canopy temperature and PAR interception in bed planted rice as compared to conventional planting. It also observed that significantly higher leaf area index of cotton under bed and furrow method as compared to flat sowing with alternate row earthing up and flat sowing with drill. Scientist observed that higher grain yield and water use efficiency of chick pea when planted on raised beds as compared to flat planting and also observed that highest yield and water use efficiency of maize when wheat straw mulch was used along with bed planting.

11.5.7 Tillage

- Tillage is the physical manipulation of soil, which is intended to destroy weeds, incorporate crop residues and amendments into soil, increase infiltration and reduce evaporation, prepare seedbed and break hard layer to facilitate root penetration.

- Tillage practices modify the crop microclimate by affecting the soil properties. Tillage increases the porosity and decreases the thermal conductivity of the soil and increases the soil temperature. The upper loose soil surface also acts as a mulch and conserves soil moisture by decreasing evaporation losses. It is observed that conventionally tilled plants intercepted 28% more sunlight during pre-bloom and 17% more light at mid-bloom before both tillage treatments reached canopy closure in cotton. Lint yield was 11%lower in the no-tillage treatment than in conventional tillage due to 8% reduction in number of bolls m^{-2}.
- It is observed that the comparable grain and straw yield of wheat with and without tillage. Whereas, other scientists observed significantly higher shelling percentage,100 seed weight and pod yield of groundnut under zero tillage as compared to conventional tillage. Also recorded higher fruit weight, length, diameter and total soluble solids in tomato under no tillage.
- It is also reported that higher yield, soil available water content, consumptive water use and water productivity under no tillage in summer as well as winter crops. As observed decrease in morning soil temperature by 0.1 to 0.4°C and increase in yield of sarson by 25% under zero tillage as compared to conventional tillage. As reported significantly higher N_2O-Nemissions under conventional tillage. Scientists reported more remunerability of the cotton–wheat system with reduced tillage in cotton and minimum tillage in wheat as compared to conventional tillage.
- It is also reported that no-till practices had lower total ecosystem respiration and evaporation, more net carbon uptake and a greater ratio of transpiration over evapotranspiration during the growing season.

11.5.8 Shelterbelts / Windbreaks

- The main function of shelterbelts is to reduce the wind speed and cold advection. Due to reduction in wind speed, they decrease the evapotranspiration and increase the temperature and moisture. Windbreaks modify the micro-climate by providing the shelter, which reduces the wind speed and turbulence. Wind breaks protect the plants from the high intensity winds. They modify the water table by reducing the water losses and hence they improve the crop productivity.
- Wind breaks also reduce the wind erosion. Scientist reported that porous windbreak modifies airflow, microclimates and hence crop yields. In dry farming, windbreak reduces crop evapotranspiration by modifying the aerodynamic component of the energy balance. As reported the positive effect of windbreaks in mitigating evapotranspiration, hence their significant applicability under dry-farming systems.

11.5.9 Cropping Systems

- Intercropping *i.e.* growing of two or more crops simultaneously on the same field such that the period of overlap is long enough to include the vegetative stage, has significant effect on microclimate and resource use efficiency. Intercropping, double cropping and other mixed cropping practices lead to enhanced efficiency of farm resources with sustainable crop production. Intercropping increases the radiation interception by the canopy and decreases its transmission towards the soil surface, thus also decreasing the soil temperature. Improved productivity thus results from either greater interception of solar radiation or higher light use efficiency or their combination. As reported more efficient radiation harvesting in intercropping of maize and beans as compared to sole crops.
- It is observed that higher PAR interception and radiation use efficiency in maize and soybean intercropping as compared to sole maize crop. It also elaborated that as a result of intercropping, microclimate within canopy can moderate temperature extremes, lower temperatures with reduced air movement leading to decreased evaporation rates and increased relative humidity, which is important in avoiding desiccation and providing favourable growth conditions even during the periods of moisture deficit. As reported soil temperature upto10°C cooler on forage under tree plantations in the tropics assisting seedling survival, soil water relations and possibly affecting the rate of litter breakdown and nitrogen mineralization. Although intercropping is also known to decrease the occurrence of weeds, insects and diseases, but modified microclimate especially moderate temperatures and increased relative humidity sometimes might become conducive for disease outbreak, especially fugal pathogens.

11.5.10 Protected Cultivation

- Protected cultivation makes it possible to obtain increased crop productivity by maintaining favourable environment for the plants. The use of netting and other type of covering has been shown to restrict air movement around the growing seedlings in higher temperature. As reported that netting effectively modified the microclimate around the growing tomato seedlings. There was an increase in daily temperature by about 3.5°C and increase in relative humidity by 10%. Elevated air temperatures inside the net along with improved moisture status and root development enhanced the uptake of nutrients such as potassium and nitrogen, thereby favouring leaf conductance and chlorophyll content.
- The net covering also offers a physical barrier reducing the occurrence of insect-pests. Also observed the delay in the infestation of cabbage by aphids under netting. Higher relative humidity recorded under netting could also affect the feeding habit of sucking pests, hence lowering their population under netting concluded that fogging system could improve the climatic conditions under

screen house and extend the growing season during adverse environmental conditions by moderating the extremes of microclimate during summer.

11.6 EFFECTS OF WEATHER ON AGRICULTURE

Agriculture, the cultivation of crops and rearing of livestock for food and commercial purposes. It is one of the pillars of the thriving economy of any country.

11.6.1 Weather on Agriculture

- Regardless of how profitable agriculture is, the prevailing weather condition of the place affects it significantly. The effect is because a change in the weather condition affects agricultural practices to thrive and yield increase. Some of these parameters include temperature, nutrient levels, soil moisture, and water availability to crops and livestock. Also, the concentration of CO_2 increases light intensity.
- Agriculture is mostly under these three headings crops, livestock, and fisheries. The climate has effects on them, though in different ways. These effects should be appropriately understood. As to know the extent of their influence on agricultural output and possible ways to curtail these effects.

11.6.2 On Crops

- In the United States and every other place in the world, a high yield of crops is responsible for abundant food supply. The effect of weather on crop yield depends on if it is an optimum year or not. Generally, weather changes that have led to extremely high temperature hurt the growth and yield of crops. A good example is in the US in the year 2010 and 2012 where high temperatures led to a decrease in crop yield.
- Extreme low temperatures could have a positive and negative impact on crops. An example is the wheat grain which becomes very vulnerable at low temperatures. However, in the same low temperatures, farmers stand to gain also. Because the soil becomes frozen such that vaporization of soil nitrogen becomes impossible.
- Hot weather conditions over an extended period could lead to drought. This drought will affect the growth of crops and the quality of rangeland to support even livestock. A good understanding of the optimum requirements of crops before planting will help enhance its yield. This understanding helps regardless of the changing weather conditions.

11.6.3 On Livestock

- In the United States, the primary source of meat is from the livestock and poultry. These sources account for the majority of cash receipts from the agricultural sectors.

- Weather conditions like a heat wave, usually affect the animals directly and even on a long-term basis. It increases their susceptibility to endemic livestock, which can decrease their milk-producing capacity. If harsh weather conditions persist for too long, there might be a loss of desirable traits.
- Cold temperature would require farmers to keep the animals warm, observing them, to avoid them from freezing to death.
- Drought due to lack of rain causes a shortage of pasture and even feed supplies. When rainfall keeps persisting livestock are always at risk to waterborne diseases which are prevalent at that time.
- When weather conditions are not predictable, it is advised by experts to adopt new veterinary practices. Also, other livestock management techniques to preserve the productivity and quality of the livestock.

11.6.4 On Fisheries

- The prevailing weather condition is regulating the aquatic ecosystem. The temperature and seasons influence the mating time and have impacts on their migration.
- Noticing results are in the life cycle of most aquatic animals like the salmon. Whenever the weather becomes hot, their susceptibility to diseases increases. Disease and other climatic factors account for the decline of salmon population in this weather condition.
- Due to the global warming and climate change, atmospheric carbon dioxide is on the increase. Atmospheric CO_2 is found to lead to the acidification of the oceans. These factors have grave consequences as it harms marine life. It can weaken their shells, displace calcium and disrupt the entire aquatic system. These disruptions affect the output of fish farming if not addressed.
- The effect weather has on the agriculture needs to be watched. Due to current climate change, farmers should be observant to understand weather trends. They will need to be flexible with agricultural practices. They should also be abreast with other ways of curtailing the adverse effects of weather to avoid food shortages.

12

CLIMATE CHANGE AND AGRICULTURE

12.1 INTRODUCTION TO CLIMATE CHANGE

- Climate change and variability are concerns of human being. Therecurrent droughts and floods threaten seriously the livelihood of billions of peoplewho depend on land for most of their needs.
- The global economy is adversely beinginfluenced very frequently due to extreme events such as droughts and floods, coldand heat waves, forest fires, landslips etc. The natural calamities like earthquakes,tsunamis and volcanic eruptions, though not related to weather disasters, maychange chemical composition of the atmosphere. It will, in turn, lead to weather related disasters.
- Increase in aerosols (atmospheric pollutants) due to emission ofgreenhouse gases such as Carbon Dioxide due to burning of fossil fuels,chlorofluorocarbons (CFCs), hydrochlorofluorocarbons (HCFCs), hydrofluorocarbons(HFCs), perfluorocarbons (PFCs) etc., Ozone depletion and UV-B filtered radiation,eruption of volcanoes, the "human hand" in deforestation in the form of forest firesand loss of wet lands are causal factors for weather extremes.
- The loss of forestcover, which normally intercepts rainfall and allows it to be absorbed by the soil,causes precipitation to reach across the land eroding top soil and causes floods anddroughts. Paradoxically, lack of trees also exacerbates drought in dry years bymaking the soil dry more quickly. Among the greenhouse gases, CO2 is thepredominant gas leading to global warming as it traps long wave radiation and emitsit back to the earth surface.
- The global warming is nothing but heating of surfaceatmosphere due to emission of greenhouse gases, thereby increasing globalatmospheric temperature over a long period of time. Such changes in surface airtemperature and consequent adverse impact on rainfall over a long period of timeare known as climate change.

If these parameters show year-to-year variations orcyclic trends, it is known as climate variability.

- However, the official definition by the United Nations Framework Convention on Climate Change (UNFCCC) is that climate change is the change that can be attributed directly or indirectly to human activity that alters the composition of the global atmosphere and which is in addition to natural climate variability observed over comparable time periods. However, scientists often use the term for any change in the climate, whether arising naturally or from human causes.
- In particular, the Intergovernmental Panel on Climate Change (IPCC) defines climate change as a change in the state of the climate that can be identified by changes in the mean and / or the variability of its properties, and that persists for an extended period, typically decades or longer.

12.2 CLIMATIC VARIABILITY

- Climate variability and change occurs when changes in Earth's climate system result in new weather patterns. Changes in climate can oscillate back and forth over years or can last for millions of years, depending on a variety of factors. In the time since the industrial revolution the climate has increasingly been affected by human activities that are causing global warming and climate change.
- The climate system receives nearly all of its energy from the sun. The climate system also radiates energy to outer space. The balance of incoming and outgoing energy, and the passage of the energy through the climate system, determines Earth's energy budget. When the incoming energy is greater than the outgoing energy, earth's energy budget is positive and the climate system is warming. If more energy goes out, the energy budget is negative and earth experiences cooling.
- The energy moving through Earth's climate system finds expression in weather, varying on geographic scales and time. Long-term averages and variability of weather in a region constitute the region's climate. Such changes can be the result of "internal variability", when natural processes inherent to the various parts of the climate system alter the distribution of energy. Examples include variability in ocean basins such as the Pacific decadal oscillation and Atlantic multidecadal oscillation. Climate variability can also result from external forcing, when events outside of the climate system's components nonetheless produce changes within the system. Examples include changes in solar output and volcanism.
- Climate variability has consequences for sea level changes, plant life, and mass extinctions; it also affects human societies.
- Climate variability is the term to describe variations in the mean state and other characteristics of climate (such as chances or possibility of extreme weather, etc.) "on all spatial and temporal scales beyond that of individual weather events. Some

of the variability does not appear to be caused systematically and occurs at random times. Such variability is called random variability or noise. On the other hand, periodic variability occurs relatively regularly and in distinct modes of variability or climate patterns.

- The term climate change is often used to refer specifically to anthropogenic climate change (also known as global warming). Anthropogenic climate change is caused by human activity, as opposed to changes in climate that may have resulted as part of Earth's natural processes. In this sense, especially in the context of environmental policy, the term climate change has become synonymous with anthropogenic global warming. Within scientific journals, global warming refers to surface temperature increases while climate change includes global warming and everything else that increasing greenhouse gas levels affect.
- A related term, climatic change, was proposed by the World Meteorological Organization (WMO) in 1966 to encompass all forms of climatic variability on time-scales longer than 10 years, but regardless of cause. During the 1970s, the term climate change replaced climatic change to focus on anthropogenic causes, as it became clear that human activities had a potential to drastically alter the climate. Climate change was incorporated in the title of the Intergovernmental Panel on Climate Change (IPCC) and the UN Framework Convention on Climate Change (UNFCCC). Climate change is now used as both a technical description of the process, as well as a noun used to describe the problem.

12.3 CAUSES

- On the broadest scale, the rate at which energy is received from the Sun and the rate at which it is lost to space determine the equilibrium temperature and climate of Earth. This energy is distributed around the globe by winds, ocean currents, and other mechanisms to affect the climates of different regions.
- Factors that can shape climate are called climate forcings or "forcing mechanisms". These include processes such as variations in solar radiation, variations in the Earth's orbit, variations in the albedo or reflectivity of the continents, atmosphere, and oceans, mountain-building and continental drift and changes in greenhouse gas concentrations. There are a variety of climate change feedbacks that can either amplify or diminish the initial forcing. Some parts of the climate system, such as the oceans and ice caps, respond more slowly in reaction to climate forcings, while others respond more quickly. There are also key threshold factors which when exceeded can produce rapid change.
- Climate variability can either occur due to external forcing or due to internal processes. Internal unforced processes often involve changes in the distribution of energy in the ocean and atmosphere, for instance changes in the thermohaline

circulation. External forcing mechanisms can be either anthropogenic (e.g. increased emissions of greenhouse gases and dust) or natural (e.g., changes in solar output, the earth's orbit, volcano eruptions).

The response of the climate system to a climate forcing might be fast (e.g., a sudden cooling due to airborne volcanic ash reflecting sunlight), slow (e.g. thermal expansion of warming ocean water), or a combination (e.g., sudden loss of albedo in the Arctic Ocean as sea ice melts, followed by more gradual thermal expansion of the water). Therefore, the climate system can respond abruptly, but the full response to forcing mechanisms might not be fully developed for centuries or even longer.

12.5 INTERNAL VARIABILITY

Climatic changes due to internal variability sometimes occur in cycles or oscillations. For other types of natural climatic change, we cannot predict when it happens; the change is called *random* or *stochastic*. From a climate perspective, the weather can be considered as being random. If there are little clouds in a particular year, there is an energy imbalance and extra heat can be absorbed by the oceans. Due to climate inertia, this signal can be 'stored' in the ocean and be expressed as variability on longer time scales than the original weather disturbances. If the weather disturbances are completely random, occurring as white noise, the inertia of glaciers or oceans can transform this into climate changes where longer-duration oscillations are also larger oscillations, a phenomenon called red noise. Many climate changes have a random aspect and a cyclical aspect. This behavior is dubbed *stochastic resonance*. Scientists generally define the five components of earth's climate system to include atmosphere, hydrosphere, cryosphere, lithosphere (restricted to the surface soils, rocks, and sediments), and biosphere.

12.6 WEATHER AND CLIMATE

- Weather is the set of meteorological conditions such as wind, rain, snow, sunshine, temperature, etc. at a particular time and place. By contrast, the term climate describes the overall long-term characteristics of the weather experienced at a place. The ecosystems, agriculture, livelihoods and settlements of a region are very dependent on its climate. The climate, therefore, can be thought of as a long-term summary of weather conditions, taking account of the average conditions as well as the variability of these conditions. The fluctuations that occur from year to year, and the statistics of extreme conditions such as severe storms or unusually hot seasons are part of the climatic variability.
- The Earth's climate has varied considerably in the past, as shown by the geological evidence of ice ages and sea level changes, and by the records of human history over many hundreds of years. The causes of past changes are not always clear but are generally known to be related to changes in ocean currents, solar activity, volcanic eruptions and other natural factors. The difference now is that global temperatures

have risen unusually rapidly over the last few decades. There is strong evidence of increase in average global air and ocean temperatures, widespread melting of snow and ice, and rising of average global sea levels. The IPCC Fourth Assessment Report concludes that the global warming is unequivocal. Atmosphere and ocean temperatures are higher than they have been at any other time during at least the past five centuries, and probably for more than a millennium. Scientists have long known that the atmosphere's greenhouse gases act as a blanket, which traps incoming solar energy and keeps the Earth's surface warmer than it otherwise would be, and that an increase in atmospheric greenhouse gases would lead to additional warming.

12.7 IMPORTANT WEATHER EXTREMES AND THEIR IMPACT AT GLOBAL LEVEL

- The year 1998 was the warmest and declared as the weather-related disaster year. It caused hurricane havoc in Central America and floods in China, India and Bangladesh. Canada and New England suffered heavily due to ice storm in January while Turkey, Argentina and Paraguay suffered with floods in June 1998. In contrast, huge crop losses were noticed in Maharashtra (India) due to un-seasonal and poor distribution of rainfall during 1997-98. The 1997/1998 *El Nino* event (The *El Nino* is nothing but warming of Pacific), the strongest of the last century, affected 110 million people and costed the global economy nearly US$ 100 billion.

- The year 2003 was the year of heat and cold waves across the world. The European Union (EU) suffered to a large extent due to heat wave that occurred in summer 2003. In India Uttar Pradesh, Bihar, West Bengal, Orissa and Andhra Pradesh are the States that experienced summer heat waves. When the EU suffered heat wave during the summer in 2003, India experienced severe cold wave from December 2002 to January 2003. Some parts of Jammu, Punjab, Haryana, Himachal Pradesh, Bihar, Uttar Pradesh and the North Eastern States experienced unprecedented cold wave. The crop yield loss varied between 10 and 100% in the case of horticultural crops and seasonal crops. The fruit size and quality were also adversely affected in horticultural crops. However, temperate fruits like apple, perch, plum and cherry gave higher yield due to extreme chilling. The damage was more in low-lying areas where cold air settled and remained for a longer time on the ground.

- High temperature in March 2004 adversely affected crops like wheat, apple, mustard, rapeseed, linseed, potato, vegetables, pea and tea across the State of Himachal Pradesh in India. The yield loss was estimated between 20% and 60% depending upon the crop. Wheat and potato harvest was advanced by 15-20 days and the flowering of apple was early by 15 days. The optimum temperature for fruit blossom and fruit set is 240 C in the case of apple while it experienced above

26 0 C for 17 days. The entire region recorded between 2.1 and 7.90 C higher maximum temperature against the normal across the State of Himachal Pradesh in March 2004 .A decline of 39% in annual cocoa yield was noticed in 2004 when compared to that of 2003 due to rise in maximum temperature of the order of 1 to 30 C from 14th January to 16th March in Central part of Kerala, India. Such trend was noticed whenever summer temperature shot up by 2 to 30 C when compared to that of normal maximum temperature of 33 to 36.50 C.

- Untimely rains and hailstorms destroyed wheat crop of 15,000 hectares (Ha.) over UP, Haryana and Punjab in Rabi season 2007 in India. In contrast, heavy snowfall over Kashmir valley was recorded in 2007 due to western disturbances. Similar was the case during monsoon 2007, causing floods across several continents (Hurricane Dean in August in Mexico) including India and Bangladesh. Heavy rains again in September in Andhra Pradesh, Karnataka and Kerala led to floods and thus the year 2007 was declared as the flood year in India. A huge crop loss was noticed in several states of the Country due to floods in kharif, 2007.
- Beijing in China had temperature as high as 160 C in February 2007, the highest since the meteorological record began in 1840, followed by one of its coldest and snowiest winter in 2008. As a result of heavy snow for a period of three weeks since 10th of January 2008, 104 million ha. of farm land was damaged in addition to destruction of 22,000 houses and the economic loss was estimated at $ 7.5 billion. The La Nina phenomena may be another reason for severe snow storms. The La Nina is a large pool of unusually cold water in the equatorial Pacific that develops over a few years and influences global weather, which is opposite to El Nino. The *El Nino* is nothing but warming of Pacific. The mercury dipped to a new low of 9.40 C over Mumbai on 6.2.2008. The frequency of such unusual weather phenomena is likely to increase across the world and huge economic loss is expected.
- The Mean Sea Level (MSL) rise is likely to be slightly less than one mm/year along the Indian coast. Sea level rise may lead to disappearance of low-lying areas of coastal belt in addition to changes in ocean biodiversity. It threatens health of corals and polar bear population. Greater number of high surges and increased occurrences of cyclones in post-monsoon period, along with increased maximum wind speed, are also expected. This phenomenon of sea level rise threatens the area of land available for farming.
- As per the United Nations Report of FAO, India stands to lose 125 million tonnes equivalent to 18% of its rainfed cereal production from climate change by 2015. China's rainfed cereal production potential of 360 million tonnes is expected to increase by 15% during the same period. It would also cause a worldwide drop in cereal crops, leaving 400 million more people at risk of hunger, and leaving three

billion people at risk of flooding and without access to fresh water supplies. The crop production losses due to climate change may also drastically increase the number of undernourished people, severely hindering progress in combating poverty and food security. The severest impact is likely to be in sub-Saharan African countries, which are least able to adapt to climate change or to compensate for it through increase in food imports. In 2004 and 2005, twenty four (24) sub- Saharan African countries faced food emergencies, caused by a lethal combination of locusts and drought. In addition, adverse hot and dry weather in United States and drought conditions in parts of the EU lowered cereal output during 2005 when compared to that of 2004. The simulation models indicate that the global warming leads to reduction in rice and wheat production in northern India.

- The Indian economy is mostly agrarian based and depends on onset of monsoon and its further behaviour. The year 2002 was a classical example to show how Indian food grains' production depends on rainfall of July and it was declared as the all-India drought, as the rainfall deficiency was 19% against the long period average of the country and 29% of the area was affected due to drought. The "AllIndia drought" is declared when the rainfall deficiency for the Country as a whole is more than 10% of normal, and when more than 20% of the Country's area is affected by drought conditions. The kharif season food grain production was adversely affected by a whopping fall of 19.1% due to "All–India drought" during monsoon 2002.

12.8 CLIMATE CHANGE AND AGRICULTURE

Based on some of the past experiences indicated above, impact of climate change on agriculture will be one of the major deciding factors influencing the future food security of mankind on the earth. Agriculture is not only sensitive to climate change but also one of the major drivers for climate change. Understanding the weather changes over a period of time and adjusting the management practices towards achieving better harvest are challenges to the growth of agricultural sector as a whole. The climate sensitivity of agriculture is uncertain, as there is regional variation in rainfall, temperature, crops and cropping systems, soils and management practices. The inter-annual variations in temperature and precipitation were much higher than the predicted changes in temperature and precipitation. The crop losses may increase if the predicted climate change increases the climate variability.

Different crops respond differently as the global warming will have a complex impact. The tropics are more dependent on agriculture as 75% of world population lives in tropics and two thirds of these people's main occupation is agriculture. With low levels of technology, wide range of pests, diseases and weeds, land degradation, unequal land distribution and rapid population growth, any impact on tropical agriculture will affect their livelihood. Rice, wheat, maize, sorghum, soybean and barley are the six major crops in the world grown in 40% cropped area, and contribute to 55% of non-meat calories and over 70% of animal feed. Consequently, any effect on these crops would adversely affect the food security.

12.9 MAIN PROJECTIONS FOR CLIMATE CHANGE AT GLOBAL LEVEL

The projections of future climate patterns are largely based on computer-based models of the climate system that incorporate the important factors and processes of the atmosphere and the oceans, including the expected growth in greenhouse gases from socio-economic scenarios for the coming decades. The IPCC has examined the published results from many different models and on the basis of the evidence has estimated that by 2100-

- The global average surface warming (surface air temperature change) will increase by 1.1 - 6.4 °C.
- The sea level will rise between 18 and 59 cm.
- The oceans will become more acidic.
- It is very likely that hot extremes, heat waves and heavy precipitation events will continue to become more frequent.
- It is very likely that there will be more precipitation at higher latitudes and it is likely that there will be less precipitation in most subtropical land areas.
- It is likely that tropical cyclones (typhoons and hurricanes) will become more intense, with larger peak wind speeds and heavier precipitation associated with ongoing increases of tropical sea surface temperatures.

12.10 LIKELY EFFECTS OF CLIMATE CHANGE ON KEY SECTORS AT GLOBAL LEVEL

The IPCC Fourth Assessment Report of the Working Group II: Impacts, Adaptation and Vulnerability describe the likely effects of climate change, including from increases in extreme events. The effects on key sectors, in the absence of countermeasures, are summarized as follows.

12.10.1 Water

Drought affected areas are likely to be more widely distributed. Heavier precipitation events are very likely to increase in frequency leading to higher flood risks. By mid-century, water availability is likely to decrease in mid-latitudes, in the dry tropics and in other regions supplied by melted water from mountain ranges. More than one sixth of the world‘s population is currently dependent on melt water from mountain ranges.

12.10.2 Food

While some mid latitude and high latitude areas will initially benefit from higher agricultural production, for many others at lower latitudes, especially in seasonally dry and tropical regions, the increases in temperature and the frequency of droughts and floods are likely to affect crop production negatively, which could increase the number of people at risk from hunger and increased levels of displacement and migration.

12.10.3 Industry, Settlement and Society

The most vulnerable industries, settlements and societies are generally those located in coastal areas and river flood plains, and those whose economies are closely linked with climate sensitive resources. This applies particularly to locations already prone to extreme weather events and especially to areas undergoing rapid urbanization. Where extreme weather events become more intense or more frequent, the economic and social costs of those events will increase.

12.10.4 Health

The projected changes in climate are likely to alter the health status of millions of people, including increased deaths, disease and injury due to heat waves, floods, storms, fires and droughts. Increased malnutrition, diarrhea disease and malaria in some areas will increase vulnerability to extreme public health, and development goals will be threatened by long term damage to health systems from disasters.

12.10.5 Projected Impact on Asia

- Asia-Pacific region may experience the worst effect on rice and wheat yields worldwide, and decreased yields could threaten the food security of 1.6 billion people in South Asia.
- The crop model indicates that in South Asia, average yields in 2050 for crops will decline from 2000 levels by about 50 percent for wheat, 17 percent for rice, and about 6 percent for maize because of climate change.
- In East Asia and the Pacific, yields in 2050 for crops will decline from 2000 levels by 20 percent for rice, 13 percent for soybean, 16 percent for wheat and 4 percent for maize because of climate change.
- With climate change, average calorie availability in Asia in 2050 is expected to be about 15 percent lower and cereal consumption is projected to decline by as much as 24 percent compared to a no-climate change scenario.
- In a no-climate change scenario, the number of malnourished children in South Asia would fall from 76 to 52 million between 2000 and 2050, and from 24 to 10 million in East Asia and the Pacific.

- Climate change will erase some of this progress, causing the number of malnourished children in 2050 to rise to 59 million in South Asia and to 14 million in East Asia and the Pacific, increasing the total number of malnourished children in Asia by about 11 million.
- To counteract the effects of climate change on nutrition, South Asia requires additional annual investments of 1.5 billion USD in rural development, and East Asia and the Pacific require almost 1 million USD more. Over half of these investments in both regions must be for irrigation expansion.
- The Asian countries most vulnerable to climate change are Afghanistan, Bangladesh, Cambodia, India, Lao PDR, Myanmar, and Nepal.
- Afghanistan, Bangladesh, India, and Nepal are particularly vulnerable to declining crop yields due to glacial melting, floods, droughts, and erratic rainfall, among other factors.

It is clear that the occurrence of floods and droughts, heat and cold waves are common across the world due to climate change. Their adverse impact on livelihood of farmers is tremendous. It is more so in India as our economy is more dependent on Agriculture. Interestingly, weather extremes of opposite in nature like cold and heat waves and floods and droughts are noticed within the same year over the same region or in different regions and likely to increase in ensuing decades. The human and crop losses are likely to be heavy. The whole climate change is associated with increasing greenhouse gases and human induced aerosols and the imbalance between them may lead to uncertainty even in year-to-year monsoon behaviour over India. Therefore, there should be a determined effort from developed and developing countries to make industrialization environment friendly by reducing greenhouse gases pumping into the atmosphere. In the same fashion, awareness programmes on climate change and its effects on various sectors viz., agriculture, health, infrastructure, water, forestry, fisheries, land and ocean biodiversity and sea level and the role played by human interventions in climate change need to be taken up on priority basis. In the process, lifestyles of people should also be changed so as not to harm earth atmosphere continuum by pumping greenhouse gases and CFCs into the atmosphere. From the agriculture point of view, effects of extreme weather events on crops are to be documented on regional scale so that it will be handy to planners in such re-occurrence events for mitigating the ill effects. Also, there is need to guide farmers on projected impact climate change and sensitise them on probable mitigation and adaptation options to minimize the risk in Agricultural sector.

- Asia is the most disaster-afflicted region in the world, accounting for about 89 percent of people affected by disasters worldwide.

12.11 OBSERVED CHANGES IN CLIMATE AND WEATHER EVENTS IN INDIA

12.11.1 Surface Temperature

At the national level, increase of 0.4° C has been observed in surface air temperatures over the past century. A warming trend has been observed along the west coast, in central India, the interior peninsula, and northeastern India. However, cooling trends have been observed in northwest India and parts of south India.

12.11.2 Rainfall

While the observed monsoon rainfall at the All India level does not show any significant trend, regional monsoon variations have been recorded. A trend of increasing monsoon seasonal rainfall has been found along the west coast, northern Andhra Pradesh, and north-western India (+10% to +12% of the normal over the last 100 years) while a trend of decreasing monsoon seasonal rainfall has been observed over eastern Madhya Pradesh, north-eastern India, and some parts of Gujarat and Kerala (-6% to - 8% of the normal over the last 100 years).

12.11.3 Extreme Weather Events

Trends of Extreme Weather Events observed in multi-decadal periods of more frequent droughts followed by less severe droughts. There has been an overall increasing trend in severe storm incidence along the coast at the rate of 0.011 events per year. While the states of West Bengal and Gujarat have reported increasing trends, a decline has been observed in Odisha. Scientists, while analysing a daily rainfall data set, have shown (i) a rising trend in the frequency of heavy rain events, and (ii) a significant decrease in the frequency of moderate events over central India from 1951 to 2000.

12.11.4 Rise in Sea Level

Using the records of coastal tide gauges in the north Indian Ocean for more than 40 years, Scientists have estimated that sea level rise was between 1.06-1.75 mm per year. These rates are consistent with 1-2 mm per year global sea level rise estimates of the IPCC.

12.11.5 Indian Summer Monsoon (ISM)

Intensity is projected to increase in the beginning of 2040 and by 10% by 2100.

12.12 IMPACTS ON HIMALAYAN GLACIERS

The Himalayas possess one of the largest resources of snow and ice and its glaciers form a source of water for the perennial rivers such as the Indus, the Ganga, and the Brahmaputra. Glacial melt may impact their long-term lean-season flows, with adverse impacts on the economy in terms of water availability and hydropower generation.

The available monitoring data on Himalayan glaciers indicates that while recession of some glaciers has occurred in some Himalayan regions in recent years, the trend is not consistent across the entire mountain chain.

12.13 SOME PROJECTIONS OF CLIMATE CHANGE OVER INDIA FOR THE 21ST CENTURY

Some modelling and other studies have projected the following changes due to increase in atmospheric GHG concentrations arising from increased global anthropogenic emissions:

12.13.1 Annual Mean Surface Temperature

The simulation studies by Indian Institute of Tropical Meteorology (IITM), Pune, estimated that annual mean surface temperature is expected to raise by the end of century, ranges from 3 to 5° C with warming more pronounced in the northern parts of India.

12.13.2 Impacts on Water Resources

Changes in key climate variables, namely temperature, precipitation and humidity, may have significant long-term implications for the quality and quantity of water. River systems of the Brahmaputra, the Ganga, and the Indus, which benefit from melting snow in the lean season, are likely to be particularly affected by the decrease in snow cover. A decline in total run-off for all river basins, except Narmada and Tapti, is projected in India's NATCOM I. A decline in run-off by more than two thirds is also anticipated for Sabarmati and Luni basins. Due to sea level rise, the fresh water sources near the coastal regions will suffer salt intrusion.

12.13.3 Impacts on Agriculture and Food Production

Food production in India is sensitive to climate changes such as variability in monsoon rainfall and temperature changes within a season. Studies by Indian Agricultural Research Institute (IARI) and others indicate greater expected loss in the Rabi crop. Every 1°C rise in temperature reduces wheat production by 4-5 Million Tonnes. Small changes in temperature and rainfall have significant effects on the quality of fruits, vegetables, tea, coffee, aromatic and medicinal plants, and basmati rice. Pathogens and insect populations are strongly dependent upon temperature and humidity, and changes in these parameters may change their population dynamics. Other impacts on agricultural and related sectors include lower yields from dairy cattle and decline in fish breeding, migration, and harvests. Global reports indicate a loss of 10-40% in crop production by 2100.

Indian climate is dominated by the southwest monsoon, which brings most of the region's precipitation. It is critical for the availability of drinking water and irrigation for agriculture. Agricultural productivity is sensitive to two broad classes of climate-induced effects (1) direct effects from changes in temperature, precipitation or carbon dioxide concentrations, and (2) indirect effects through changes in soil moisture and the distribution and frequency of infestation by pests and diseases. Rice and wheat yields could decline considerably with climatic changes (IPCC 1996; 2001). However, the vulnerability of agricultural production to climate change depends not only on the physiological response of the affected plant, but also on the ability of the affected socio-economic systems of production to cope with changes in yield, as well as with changes in the frequency of droughts or floods. The adaptability of farmers in India is severely restricted by the heavy reliance on natural factors and the lack of complementary inputs and institutional support systems. The loss in net revenue at the farm level is estimated to range between 9% and 25% for a temperature rise of 2 °C to 3.5 °C. Scientists also estimated that a 2°C rise in mean temperature and a 7% increase in mean precipitation would reduce net revenues by 12.3% for the country as a whole. Agriculture in the coastal regions of Gujarat, Maharashtra, and Karnataka is found to be the most negatively affected. Small losses are also indicated for the major food-grain producing regions of Punjab, Haryana, and western Uttar Pradesh. On the other hand, West Bengal, Orissa, and Andhra Pradesh are predicted to benefit to a small extent from warming.

12.13.4 Impacts on Health

- Changes in climate may alter the distribution of important vector species (for example, malarial mosquitoes) and may increase the spread of such diseases to new areas. If there is an increase of 3.8 °C in temperature and a 7% increase in relative humidity, the transmission windows i.e., months during which mosquitoes are active, will be open for all 12 months in 9 states in India.
- The transmission windows in Jammu and Kashmir and in Rajasthan may increase by 3-5 months. However, in Orissa and some southern states, a further increase in temperature is likely to shorten the transmission window by 2-3 months.

12.13.5 Impacts on Forests

Climate projections indicate that the country is likely to experience shift in forest types, with consequent changes in forests produce, and, in turn, livelihoods based on those products. Correspondingly, the associated biodiversity is likely to be adversely impacted.

12.13.6 Vulnerability to Extreme Events

- Heavily populated regions such as coastal areas are exposed to climatic events such as cyclones, floods, drought, and large declines in sown areas in arid and semiarid zones occur during climate extremes. Large areas in Rajasthan, Andhra Pradesh, Gujarat, and Maharashtra and comparatively small areas in Karnataka, Orissa, Madhya Pradesh, Tamil Nadu, Bihar, West Bengal, and Uttar Pradesh are frequented by drought.
- About 40 million hectares of land is flood-prone, including most of the river basins in the north and the northeastern belt, affecting about 30 million people on an average each year. Such vulnerable regions may be particularly impacted by climate change.

12.13.7 Impacts on Coastal Areas

- A mean Sea Level Rise (SLR) of 15-38 cm is projected along India's coast by the mid 21st century and of 46-59 cm by 2100.
- In addition, a projected increase in the intensity of tropical cyclones poses a threat to the heavily populated coastal zones in the country (NATCOM, 2004).

12.13.8 Impacts on Biodiversity

The Intergovernmental Panel on Climate Change has projected that global average temperature increase during 21st century will range from 1.4o to 4o Celsius. Research by the Consultative Group on International Agricultural Research based on distribution models of wild relatives of three staple crops of the poor i.e. Peanuts, cowpea and potato suggests that 16-22 per cent of wild species will be threatened by extinction by 2055. Loss of genetic diversity can have serious long-term consequences globally.

12.13.9 Impacts on Pest

- Some of the most dramatic effects of climate change on pests and diseases are likely to be seen among arthropod insects like mosquitoes, midges, ticks, fleas and sand flies, and the viruses they carry. With changes in temperature and humidity levels, the populations of these insects may expand their geographic range, and expose animals and humans to diseases to which they have no natural immunity. Plant pests, which include insects, pathogens and weeds, continue to be one of the biggest constraints to food and agricultural production. Fruit flies, for instance, cause extensive damage to fruits and vegetables production.
- Controlling such pests often requires the use of pesticides, which can have serious side effects on human health and the environment. Climate change may also play a role in food safety. A growing number of pests and diseases could lead to higher and even unsafe levels of pesticide residue and veterinary drugs in local food supplies. And changes in rainfall, temperature and relative humidity can readily contaminate foods like groundnuts, wheat, maize, rice and coffee with fungi that produce potentially fatal mycotoxins.

12.13.10 Effect of Rising Temperature

- In temperate climate, increased temperature could increase insect population. Rising temperature may affect insect survival, development, geographic range and population size. It may affect insect physiology. Under such situation some insects take several years to complete life cycle (Cicadas, Arctic moths) and some insects develop quickly at certain temperature range based on degree days (cabbage maggot, onion maggot, European corn borer, Colorado potato beetle, aphids, diamond back moth). Therefore, crop damage increase. Migratory pests may migrate earlier. Natural enemy-host relationship may affect resulting into reduced parasitism. Rising temperature may change gender ratios of insects such as thrips. The population of insects will increase due to lower winter mortality of insects as a result of warmer winter.

- Higher temperature may tend to shift crops geographically and hence its pests to higher altitudes. From fossil records it is understood that diversity of insect species and intensity of feeding increase with increase in temperature. Increased temperature could decrease insect population (aphids) in some crops, which cannot be grown in higher temperature. The same condition may be conducive for increased activity of natural enemies of that pest further reducing its population.

12.13.11 Effect of Precipitation

- Rain drops physically dislodge the insects from their hosts such as leafhoppers, plant hoppers, thrips, cut worms etc. while others drown to death e.g. mealy bugs, pupae of fruit fly, Helicoverpa, Spodoptera, Etiella , rice stem borers etc. Flooding is used as a control measure for termites and stem borers too.
- Heavy rainfall causes pest epizootics by fungal pathogens (sugarcane pyrilla). It is anticipated that cutworm infestation will be more in future because they are sensitive to flooding and summer rainfall, which will increase in future.

12.13.12 Effect of Rising CO_2 Level

Carbon dioxide is a perfect example of a change that could have both positive and negative effects. Carbon dioxide is expected to have positive physiological effects through increased photosynthesis. The impact is higher on C3 crops such as wheat and rice than on C4 plants like maize and grasses. The direct effects of changes in CO_2 concentration will be through changes in temperature, precipitation and radiation. However, indirect effects will bring changes in soil moisture and infestation by pests and diseases because of rising temperature and relative humidity. Such indirect effects through the increase in temperature will reduce crop duration, increase crop respiration rates, evapo-transpiration, decrease fertilizer use efficiency and enhance pest infestation. There is general consensus that the yield of main season (Kharif) crop will increase due to the effect of higher CO_2 levels. However, large yield decreases are predicted for the rabi crops because of increased temperatures. The rising CO_2 level in atmosphere has indirect impact on insect population.

Soybean crop in higher CO_2 concentration had 57% more insect damage (Japanese beetle, Leafhopper, Root worm, Mexican bean beetle) than earlier. It causes increase in level of simple sugars in the leaves that stimulates more feeding by insects. Increased C/N ratio in plant tissue due to increased CO_2 level may slow insect development and increase life stages of insect pests vulnerable to attack by parasitoids. At our current rate of green house emissions, several of the main pests that target corn will increase in number and expand their range by the end of 21st century.

12.13.13 Effect on insecticide Use Efficiency

- Warmer temperature requires more number of insecticide applications (i.e., three, more than normal) for controlling corn pests. Entomologists predict more generation of insects in warm climate that necessitates more number of insecticide applications. It will increase cost of protection and environmental pollution.
- Synthetic pyrethroids and naturalite spinosad will be less effective in higher temperature. Therefore, it is advisable for the farmers not to use insecticides with similar mode of action frequently to avoid development of resistance in case of more number of applications. Cultural management practices e.g. early planting may not be helpful because of early emergence of pests due to warmness.

12.13.14 Effect on Natural Pest Control

- Global warming is expected to make regional climates more varied and unpredictable which could affect relationship between insects and their natural enemies. In years of most variable rainfall, the caterpillars have significantly less number of parasitoids. This could be because the parasitoids use cues e.g. change in local climate to determine the best time for laying eggs.
- Unpredictable rains might disrupt the parasitoids ability to track their caterpillar hosts. The wasps use start of the rain as cues to hatch out of their cocoons and look for a caterpillar to lay their eggs. If the rains are late, they emerge late and may not find larval stage of host resulting in reduced natural pest control. Due to changes in climate, the frequency of occurrence of droughts, heat waves, windstorms and floods etc. will increase disrupting the natural ecosystems.

12.13.15 Effect on Forest Insect Pests

- It is difficult to predict impacts of climate change on forest insect pests because of complexity of interactions between insects and trees. Population of green spruce aphid (*Elatobium abietinum*) will increase due to global warming. The spruce bark beetle (*Dendroctonus micans*) will increase due to warming because its predator *Rhizophagus grandis* is benefited by temperature rise. The Asian long horn beetle (*Anoplophora glabripennis*) population will increase in warmer coastal areas that attack street plantations.
- In general, it is assumed that many forest insect pests will increase as a result of climate change. At the same time, it is likely that pests' natural enemies will benefit. So it is unclear to some extent as to what the overall effect of global climate change will be on forest insect pests.

12.14 IMPACT OF CLIMATE CHANGE ON DISEASE

Any direct yield gains caused by increased CO_2 or climate change could be offset partly or entirely by losses caused by phytophagous insects, plant pathogens and weeds. It is, therefore, important to consider these biotic constraints on crop yields under climate change.

12.14.1 Impacts on Plant Pathosystems

Climate change has the potential to modify host physiology and resistance and to alter stages and rates of development of the pathogen. The most likely impacts would be shifts in the geographical distribution of host and pathogen, changes in the physiology of host-pathogen interactions and changes in crop loss. Another important impact may be through changes in the efficacy of control strategies.

12.14.2 Geographical Distribution of Host and Pathogen

New disease complexes may arise and some diseases may cease to be economically important if warming causes a pole ward shift of agro climatic zones and host plants migrate into new regions. Pathogens would follow the migrating hosts and may infect remnant vegetation of natural plant communities not previously exposed to the often more aggressive strains from agricultural crops. The mechanism of pathogen dispersal, suitability of the environment for dispersal, survival between seasons, and any change in host physiology and ecology in the new environment will largely determine how quickly pathogens become established in a new region. Changes may occur in the type, amount and relative importance of pathogens and affect the spectrum of diseases affecting a particular crop. This would be more pronounced for pathogens with alternate hosts. Plants growing in marginal climate could experience chronic stress that would predispose them to insect and disease outbreaks. Warming and other changes could also make plants more vulnerable to damage from pathogens that are currently not important because of unfavorable climate.

12.14.3 Physiology of Host-Pathogen Interactions – Elevated CO_2

Increases in leaf area and duration, leaf thickness, branching, tillering, stem and root length and dry weight are well known effects of increased CO_2 on many plants. Scientists have suggested that elevated CO_2 would increase canopy size and density, resulting in a greater biomass of high nutritional quality. When combined with increased canopy humidity, this is likely to promote foliar diseases such as rusts, powdery mildews, leaf spots and blights. The decomposition of plant litter is an important factor in nutrient cycling and in the saprophytic survival of many pathogens. Increased C:N ratio of litter is a consequence of plant growth under elevated CO_2. Scientists have indicated that decomposition of high-CO_2 litter occurs at a slower rate. Increased plant biomass, slower decomposition of litter and higher winter temperature could increase pathogen survival on over wintering crop residues and increase the amount of initial inoculum available to infect subsequent crops. Host pathogen interactions in selected fungal pathosystems, two important trends have emerged on the effects of elevated CO_2. First, the initial establishment of the pathogen may be delayed because of modifications in pathogen aggressiveness and/or host susceptibility. The second important finding has been an increase in the fecundity of pathogens under elevated CO_2.

12.14.4 Elevated Temperature

Increases in temperature can modify host physiology and resistance. Considerable information is available on heat-induced susceptibility and temperature-sensitive genes. In contrast, lignification of cell walls increased in forage species at higher temperatures to enhance resistance to fungal pathogens. Impacts would, therefore, depend on the nature of the host-pathogen interactions and the mechanism of resistance. Agricultural crops and plants in natural communities may harbor pathogens as symptomless carriers, and disease may develop if plants are stressed in a warmer climate. Host stress is an especially important factor in decline of various forest species.

12.14.5 Crop Loss

At elevated CO_2, increased partitioning of assimilates to roots occurs consistently in crops such as carrot, sugar beet and radish. If more carbon is stored in roots, losses from soil borne diseases of root crops may be reduced under climate change. In contrast, for foliar diseases favored by high temperature and humidity, increases in temperature and precipitation under climate change may result in increased crop loss. The effects of enlarged plant canopies from elevated CO_2 could further increase crop losses from foliar pathogens.

12.15 CURRENT ACTIONS FOR ADAPTATION AND MITIGATION IN INDIA

- Adaptation, in the context of climate change, comprises the measures taken to minimize the adverse impacts of climate change, e.g. relocating the communities living close to the sea shore, for instance, to cope with the rising sea level or switching to crops that can withstand higher temperatures.
- Mitigation comprises measures to reduce the emissions of greenhouse gases that cause climate change in the first place, e.g. by switching to renewable sources of energy such as solar energy or wind energy or nuclear energy instead of burning fossil fuel in thermal power stations.
- Current government expenditure in India on adaptation to climate variability exceeds 2.6% of the GDP, with agriculture, water resources, health and sanitation, forests, coastal-zone infrastructure and extreme weather events, being specific areas of concern.

12.15.1 Programmes

- **Crop Improvement**: Programmes address measures such as development of aridland crops and pest management as well as capacity building of extension workers and NGOs to support better vulnerability-reducing practices.

- **Drought Proofing:** The current programmes seek to minimize the adverse effects of drought on production of crops and livestock, and on productivity of land, water and human resources, so as to ultimately lead to drought proofing of the affected areas. They also aim to promote overall economic development and improve the socio-economic conditions of the resource poor and disadvantaged.
- **Forestry:** India has a strong and rapidly growing afforestation programme. The afforestation process was accelerated by the enactment of the Forest Conservation Act of 1980, which aimed at stopping the clearing and degradation of forests through a strict, centralized control of the rights to use forestland and mandatory requirements of compensatory afforestation in case of any diversion of forestland for any non-forestry purpose. In addition, an aggressive afforestation and sustainable forest management programme resulted in annual reforestation of 1.78 mha during 1985-1997, and is currently 1.1 mha annually. Due to this, the carbon stocks in Indian forests have increased over the last 20 years to 9 -10 gigatons of carbon (GtC) during 1986 to 2005.
- **Water**: The National Water Policy (2002) stresses that non-conventional methods for utilization of water, including inter-basin transfers, artificial recharge of groundwater, and desalination of brackish or sea water, as well as traditional water conservation practices like rainwater harvesting, including roof-top rainwater harvesting, should be practiced to increase the utilizable water resources. Many states now have mandatory water harvesting programmes in several cities.
- **Coastal Regions**: In coastal regions, restrictions have been imposed in the area between 200 m and 500 m of the HTL (High Tide Line) while special restrictions have been imposed in the area up to 200 m to protect the sensitive coastal ecosystems and prevent their exploitation. This, simultaneously, addresses the concerns of the coastal population and their livelihood. Some specific measures taken in this regard include construction of coastal protection infrastructure and cyclone shelters, as well as plantation of coastal forests and mangroves.
- **Risk Financing**: Two risk-financing programmes support adaptation to climate impacts. The Crop Insurance Scheme supports the insurance of farmers against climate risks, and the Credit Support Mechanism facilitates the extension of credit to farmers, especially for crop failure due to climate variability.
- **Disaster Management**: The National Disaster Management programme provides grants-in-aid to victims of weather related disasters and manages disaster relief operations. It also supports proactive disaster prevention programmes, including dissemination of information and training of disaster management staff.

12.16 INDIA'S POLICY STRUCTURE RELEVANT TO GHG MITIGATION

- India has in place a detailed policy, regulatory and legislative structure that relates strongly to GHG mitigation: The Integrated Energy Policy was adopted in 2006. Some of its key provisions are:
  - Promotion of energy efficiency in all sectors
  - Emphasis on mass transport
  - Emphasis on renewable including biofuels plantations
  - Accelerated development of nuclear and hydropower for clean energy
  - Focused R & D on several clean energy related technologies
- The experience gained so far enables India to embark on an even more proactive approach through National Action Plan on Climate Change (NAPCC). NAPCC identifies measures that promote our development objectives while also yielding co-benefits for addressing climate change effectively. It outlines a number of steps to simultaneously advance India's development and climate change related objectives of adaptation and mitigation.
- The following eight National Missions form the core of the National Action Plan, representing multi-pronged, long-term and integrated strategies for achieving key goals in the context of climate change:
  - National Solar Mission
  - National Mission for Enhanced Energy Efficiency
  - National Mission on Sustainable Habitat
  - National Water Mission
  - National Mission for Sustaining the Himalayan Ecosystem
  - National Mission for a "Green India"
  - National Mission for Sustainable Agriculture
  - National Mission on Strategic Knowledge for Climate Change

12.17 COPING OPTIONS FOR FARMERS

12.17.1 Awareness on Climate Change:

Farmers need to be sensitized on climate variability, climate change, its impact on crop production, and coping options.

12.17.2 Agromet Advisories:

The farming community is provided with Agromet advisories. Its bulletins are prepared taking into account the prevailing weather, soil and crop condition and weather prediction.

Measures / practices / suggestions to be taken in view of weather forecast to minimize the losses and optimize inputs (Land preparation, selection of crop & cultivars, Date of sowing, Date of harvesting, Irrigation scheduling, Pesticides & Fertilizer application, Extreme weather events, etc.). Agro-advisory bulletin consist of three parts (i) Weather events occurred during past week and weather forecast for five days ahead. These forecast includes weather parameters like cloud amount, rainfall, average Wind Speed, Wind Direction, RH, maximum and minimum temperature. (ii) It contains actual information on state and stage of crop growth, ongoing agricultural operations, disease and insect pest occurrence. (iii) It provides value added information on various farm activities to be taken based on weather.

An Insurance Product based on a Weather Index: The basic idea of weather insurance is to estimate the percentage deviation in crop output due to adverse weather conditions. Weather insurance protects against additional expenses or loss of profit from specific bad weather events.

- **Contingency Planning:** It is a plan devised for an exceptional risk, which is impractical or impossible to avoid. Ministry of Agriculture, through Indian Council of Agricultural Research (ICAR) and State Agricultural Universities (SAUs), is working on district-specific contingency plan for the agriculture and allied sectors. This includes fisheries and animal husbandry and started in March 2010 under Rashtriya Krishi Vikas Yojana (RKVY). Contingency plans for about 300 out of the total 600 odd districts were prepared and validated by experts and hosted in Department of Agriculture and Cooperation (DAC), ICAR, Centre for Research in Dry land Areas (CRIDA) website. The comprehensive district-specific document is having details on the crops and cultivation practices to be adopted in case of deficient or delay in monsoon, unseasonal rains, frosts or unusually high temperature, excessive rains etc. Each district is a scientific document for adaptation in case of eventualities.
- **Demonstration of Climate Resilient Technologies to the farmers may be undertaken by the Extension personnel in the areas of:**
    - Natural resource management: Interventions on in-situ moisture conservation, rain water harvesting and recycling for supplemental irrigation, improved drainage in flood prone areas, conservation tillage, ground water recharge and water saving irrigation methods etc.
    - Crop Production: Introducing drought/temperature tolerant varieties, advancement of planting dates of rabi crops in areas with terminal heat stress, water saving paddy cultivation methods (SRI, aerobic, direct seeding), frost management in horticulture through fumigation, community nurseries for delayed monsoon, custom hiring centers for farm machineries for timely

planting, location specific intercropping systems with high sustainable yield index etc

- Livestock and Fisheries: Augmentation of fodder production during droughts/floods, improving productivity of Common Property Resources (CPRs), promotion of improved fodder/feed storage methods, preventive vaccination, improved shelters for reducing heat/cold stress, management of fish ponds/tanks during water scarcity and flooding etc.
- Institutional Interventions: Institutional interventions, either by strengthening the existing ones or initiating new ones, relating to seed bank, fodder bank, custom hiring center, collective marketing, and introduction of weather index based insurance and climate literacy through a village level weather station.

13

WIND AND AGROMETEOROLOGY

13.1 INTRODUCTION

Wind is the movement of air caused by the uneven heating of the Earth by the sun. It does not have much substance—you cannot see it or hold it—but you can feel its force. It can dry your clothes in summer and chill you to the bone in winter. It is strong enough to carry sailing ships across the ocean and rip huge trees from the ground. It is the great equalizer of the atmosphere, transporting heat, moisture, pollutants, and dust great distances around the globe. Landforms, processes, and impacts of wind are called Aeolian landforms, processes, and impacts. Differences in atmospheric pressure generate winds. At the Equator, the sun warms the water and land more than it does the rest of the globe. Warm equatorial air rises higher into the atmosphere and migrates toward the poles. This is a low-pressure system. At the same time, cooler, denser air moves over Earth's surface toward the Equator to replace the heated air. This is a high-pressure system. Winds generally blow from high-pressure areas to low-pressure areas. The boundary between these two areas is called a front. The complex relationships between fronts cause different types of wind and weather patterns.

Prevailing winds are winds that blow from a single direction over a specific area of the Earth. Areas where prevailing winds meet are called convergence zones. Generally, prevailing winds blow east-west rather than north-south. This happens because Earth's rotation generates what is known as the Coriolis effect. The Coriolis effect makes wind systems twist counter-clockwise in the Northern Hemisphere and clockwise in the Southern Hemisphere.

The Coriolis effect causes some winds to travel along the edges of the high-pressure and low-pressure systems. These are called geostrophic winds. In 1857, Dutch meteorologist Christoph Buys Ballot formulated a law about geostrophic winds: When you stand with your back to the wind in the Northern Hemisphere, low pressure is always to your left. (In the Southern Hemisphere, low-pressure systems will be on your right.)

13.2 WIND ZONES

The Earth contains five major wind zones: polar easterlies, westerlies, horse latitudes, trade winds, and the doldrums.

13.2.1 Polar Easterlies

Polar easterlies are dry, cold prevailing winds that blow from the east. They emanate from the polar highs, areas of high pressure around the North and South Poles. Polar easterlies flow to low-pressure areas in sub-polar regions.

13.2.2 Westerlies

Westerlies are prevailing winds that blow from the west at mid latitudes. They are fed by polar easterlies and winds from the high-pressure horse latitudes, which sandwich them on either side. Westerlies are strongest in the winter, when pressure over the pole is low, and weakest in summer, when the polar high creates stronger polar easterlies. The strongest westerlies blow through the "Roaring Forties," a wind zone between 40 and 50 degrees latitude in the Southern Hemisphere. Throughout the Roaring Forties, there are few landmasses to slow winds. The tip of South America and Australia, as well as the islands of New Zealand, are the only large landmasses to penetrate the Roaring Forties. The westerlies of the Roaring Forties were very important to sailors during the Age of Exploration, when explorers and traders from Europe and western Asia used the strong winds to reach the spice markets of Southeast Asia and Australia.

Westerlies have an enormous impact on ocean currents, especially in the Southern Hemisphere. Driven by westerlies, the powerful Antarctic Circumpolar Current (ACC) rushes around the continent (from west to east) at about 4 kilometers per hour (2.5 miles per hour). In fact, another name for the Antarctic Circumpolar Current is the West Wind Drift. The ACC is the largest ocean current in the world, and is responsible for transporting enormous volumes of cold, nutrient-rich water to the ocean, creating healthy marine ecosystems and food webs.

13.2.3 Horse Latitudes

The horse latitudes are a narrow zone of warm, dry climates between westerlies and the trade winds. Horse latitudes are about 30 and 35 degrees north and south. Many deserts, from the rainless Atacama of South America to the arid Kalahari of Africa, are part of the horse latitudes.

The prevailing winds at the horse latitudes vary, but are usually light. Even strong winds are often short in duration.

13.3 TRADE WINDS

Trade winds are the powerful prevailing winds that blow from the east across the tropics. Trade winds are generally very predictable. They have been instrumental in the history of exploration, communication, and trade. Ships relied on trade winds to establish quick, reliable routes across the vast Atlantic and, later, Pacific Oceans. Even today, shipping depends on trade winds and the ocean currents they drive.

In 1947, Norwegian explorer Thor Hyerdahl and a small crew used trade winds to travel from the coast of Peru to the coral reefs of French Polynesia, more than 6,920 kilometers (4,300 miles), in a sail-powered raft. The expedition, named after the raft (Kon-Tiki) aimed to prove that ancient mariners could have used predictable trade winds to explore wide stretches of the Pacific.

Trade winds that form over land (called continental trade winds) are warmer and drier than those that form over the ocean (maritime trade winds). The relationship between continental and maritime trade winds can be violent.

Most tropical storms, including hurricanes, cyclones, and typhoons, develop as trade winds. Differences in air pressure over the ocean cause these storms to develop. As the dense, moist winds of the storm encounter the drier winds of the coast, the storm can increase in intensity. Strong trade winds are associated with a lack of precipitation, while weak trade winds carry rainfall far inland. The most famous rain pattern in the world, the Southeast Asian monsoon, is a seasonal, moisture-laden trade wind. Besides ships and rainfall, trade winds can also carry particles of dust and sand for thousands of kilometers. Particles from Saharan sand and dust storms can blow across islands in the Caribbean Sea and the U.S. state of Florida, more than 8,047 kilometers (5,000 miles) away. Dust storms in the tropics can be devastating for the local community. Valuable topsoil is blown away and visibility can drop to almost zero. Across the ocean, dust makes the sky hazy. These dust storms are often associated with dry, low-pressure areas and a lack of tropical storms.

Doldrums

The place where trade winds of the two hemispheres meet is called the intertropical convergencezone(ITCZ).TheareaaroundtheITCZiscalledthe doldrums. Prevailing winds in the doldrums are very weak, and the weather is unusually calm. The ITCZ straddles the Equator. In fact, the low-pressure doldrums are created as the sun heats the equatorial region and causes air masses to rise and travel north and south. (This warm, low-pressure equatorial wind descends again around the horse latitudes. Some equatorial air masses return to the doldrums as trade winds, while others circulate in the other direction as westerlies.) Although monsoons impact tropical as well as equatorial regions, the wind itself is created as the ITCZ moves slightly away from the Equator each season. This change in the doldrums disturbs the usual air pressure, creating the moisture-laden Southeast Asian monsoon.

13.4 RESULTS OF WIND

Wind traveling at different speeds, different altitudes, and over water or land can cause different types of patterns and storms.

13.4.1 Jet Streams

Jet streams are geostrophic winds that form near the boundaries of air masses with different temperatures and humidity. The rotation of the Earth and its uneven heating by the sun also contribute to the formation of high-altitude jet streams. These strong, fast winds in the upper atmosphere can blow 480 kph (298 mph). Jet streams blow through a layer of the atmosphere called the stratosphere, at altitudes of 8 to 14 kilometers (5 to 9 miles) above Earth's surface. There is little turbulence in the stratosphere, which is why commercial airline pilots like to fly in this layer. Riding with jet streams saves time and fuel. Have you ever heard someone talk about a headwind or tailwind when they are talking about airplanes? These are jet streams. If they are behind the plane, pushing it forward, they are called tailwinds. They can help you get to your destination more quickly. If the winds are in front of the plane, pushing it back, they are called headwinds. Strong headwinds can cause flight delays.

13.4.2 Hurricane

A hurricane is a giant, spiraling tropical storm that can pack wind speeds of over 257 kph (160 mph) and unleash more than 9 trillion liters (2.4 trillion gallons) of rain. These same tropical storms are known as hurricanes in the Atlantic Ocean, cyclones in the northern Indian Ocean, and typhoons in the western Pacific Ocean. These tropical storms have a spiral shape. The spiral (swirling counter-clockwise in the Northern Hemisphere and clockwise in the Southern Hemisphere) develops as a high-pressure area twists around a low-pressure area. The Atlantic Ocean's hurricane season peaks from mid-August to late October and averages five to six hurricanes per year. Wind conditions that can lead to hurricanes are called tropical disturbances. They begin in warm ocean waters when the surface temperatures are at least 26.6 degrees Celsius (80 degrees Fahrenheit). If the disturbance lasts for more than 24 hours and gets to speeds of 61 kph (38 mph), it becomes known as a tropical depression. When a tropical depression speeds up to 63-117 kph (39-73 mph), it is known as a tropical storm, and is given a name. Meteorologists name the storms in alphabetical order, and alternate with female and male names. When a storm reaches 119 kph (74 mph), it becomes a hurricane and is rated from 1 to 5 in severity on the Saffir Simpson scale. A Category 5 hurricane is the strongest storm possible on the Saffir-Simpson scale. Winds of a Category 5 blow at 252 kph (157 mph). Hurricanes spin around a low-pressure (warm) center known as the "eye." Sinking air inside the eye makes it very calm. The eye is surrounded by a violent circular "eye wall." This is where the storm's strongest winds and rain are. Hurricane Ethel, the strongest hurricane in recorded history, roared across the Gulf of Mexico in September 1960. Winds were sustained at 260 kph

(160 mph). However, Hurricane Ethel quickly dissipated. Although its winds ultimately blew as far north as the U.S. states of Ohio and Kentucky, by the time it hit the coastline of the U.S. states of Louisiana and Mississippi, the storm surge was only about 1.5 meters (5 feet). Only one person died as a result of Hurricane Ethel, and damage to buildings and boats was limited to less than $2 million.

Hurricanes bring destruction to coastal ecosystems and communities. When a hurricane reaches land, it often produces waves that can reach 6 meters (20 feet) high and be pushed by high winds 161 kilometers (100 miles) inland. These storm surges are extremely dangerous and cause 90 percent of all hurricane deaths. The deadliest hurricane on record is the Great Hurricane of 1780. Although sophisticated meteorological equipment was not available at that time, winds may have reached 320 kph (200 mph) as the hurricane hit Barbados and other islands in the Caribbean Sea. This may have been enough to strip the bark from trees. More than 20,000 people died as a result of the hurricane as it made its way across Barbados, St. Lucia, Martinique, Dominica, Guadeloupe, Dominican Republic, Bahamas, Turks and Caicos, and Bermuda. Although it decreased in intensity, the hurricane was tracked through the U.S. state of Florida before dissipating in the Canadian province of Newfoundland. Hurricanes can be destructive in other ways. High winds can create tornadoes. Heavy rains contribute to floods and landslides, which may occur many kilometers inland. Damage to homes, businesses, schools, hospitals, roads, and transportation systems can devastate communities and entire regions. Hurricane Katrina, which blew through the Gulf of Mexico and into the southern U.S. in 2005, is the most expensive hurricane in recorded history. Damage to buildings, vehicles, roads, and shipping facilities is estimated at about $133.8 billion (adjusted for inflation). New Orleans, Louisiana, was almost completely devastated by Hurricane Katrina. New Orleans, as well as Mobile, Alabama, and Gulfport, Mississippi, took years to recover from the damage done to their structures and infrastructure. The best defense against a hurricane is an accurate forecast that gives people time to get out of its way. The National Hurricane Center issues hurricane watches for storms that may endanger communities, and hurricane warnings for storms that will reach land within 24 hours.

13.4.3 Cyclones

Cyclones blow through the Indian Ocean in the same way hurricanes blow across the Atlantic. Cyclones blow in with air masses from the east, often the South China Sea, or the south.

The most powerful and devastating cyclone in recorded history was the 1970 Bhola Cyclone. Like Hurricane Katrina, the Bhola Cyclone was a Category 3 storm. Its winds were about 185 kph (115 mph) as it made landfall along the coast of the Bay of Bengal, in what is today Bangladesh. More than 300,000 people died, and more than a million were made homeless. Cyclone winds devastated fishing villages, and storm surges drowned crops. Economic damage from the Bhola Cyclone was more than $479 million, adjusted for inflation.

13.4.4 Typhoon

Typhoons are tropical storms that develop over the northwest Pacific Ocean. Their formation is identical to hurricanes and cyclones. Typhoons form as equatorial winds and blow westward before turning north and merging with westerlies around the mid-latitudes. Typhoons can impact a wide area of the eastern Pacific. The islands of the Philippines, China, Vietnam, and Japan are the most affected. However, typhoons have also been recorded as far as the U.S. states of Hawaii and even Alaska. Typhoons are often associated with extremely heavy rainfall. The wettest typhoon ever recorded was Typhoon Morakot in 2009. Morakot devastated the entire island of Taiwan, with winds of about 140 kph (85 mph). Storm surges and floods caused by those winds, however, caused the most damage. More than 277 centimeters (109 inches) of rain drenched Taiwan, leading to 461 deaths and $6.2 billion in damage.

13.4.5 Nor'easters and Blizzards

A nor'easter is a strong winter storm combining heavy snowfall, strong winds, and very cold temperatures. It blows from the northeast along the East Coast of the U.S. and Canada. A strong nor'easter is called a blizzard.

The U.S. Weather Service calls a storm a blizzard when the storm has wind speeds of more than 56 kph (35 mph) and low visibility. (Visibility is the distance that a person can see—blizzards, like fog, make visibility difficult and a task like driving dangerous.) The storm must go on for a prolonged period of time to be classified as a blizzard, usually a few hours. Blizzards can isolate and paralyze areas for days, especially if the area rarely has snowfall and does not have the equipment to clear it from the streets. The Great Blizzard of 1888 was perhaps the worst in U.S. recorded history. Winds of up to 72 kph (45 mph) whipped the East Coast from Chesapeake Bay to as far north as Nova Scotia, Canada. More than 147 centimeters (58 inches) of snow fell across the region, causing freezing temperatures and massive flooding as the snow melted. The Great Blizzard resulted in 400 deaths and $1.2 billion in damage.

Monsoon

A monsoon is a seasonal change in the prevailing wind system of an area. They always blow from cold, high-pressure regions. Monsoons are part of a yearlong cycle of uneven heating and cooling of tropical and mid-latitude coastal regions. Monsoons are part of the climate of Australia, Southeast Asia, and in the southwestern region of North America. The air over land is heated and cooled more quickly than the air over the ocean. During summer, this means warm land-air rises, creating a space for the cool and moist air from the ocean. As the land heats the moist air, it rises, cools, condenses, and falls back to Earth as rain. During the winter, land cools more quickly than the ocean. The warm air over the ocean rises, allowing cool land-air to flow in. Most winter monsoons are cool and dry, while summer monsoons are warm and moist. Asia's winter monsoons bring cool, dry air from the Himalaya mountains. The famous summer monsoon, on the other hand, develops over the Indian Ocean, absorbing tremendous amounts of moisture. Summer monsoons bring warmth and precipitation to India, Sri Lanka, Bangladesh, and Myanmar. The summer monsoon is essential for the health and economies of the Indian subcontinent. Aquifers are filled, allowing water for drinking, hygiene, industry, and irrigation.

Tornado

A tornado, also called a twister, is a violently rotating funnel of air. Tornadoes can occur individually or in multiples, as two spinning vortexes of air rotating around each other. Tornadoes can occur as waterspouts or landspouts, spinning from hundreds of meters in the air to connect the land or water with clouds above. Although destructive tornadoes can occur at any time of day, most of them occur between 4 and 9 p.m. local time. Tornadoes often occur during intense thunderstorms called supercells. A supercell is a thunderstorm with a powerful, rotating updraft. (A draft is simply a vertical movement of air.) This powerful updraft is called a mesocyclone. A mesocyclone contains rotating drafts of air 1 to 10 kilometers (1 to 6 miles) in the atmosphere. When rainfall increases in the supercell, rain can drag the mesocyclones down with it to the ground. This downdraft is a tornado. Depending on the temperature and moisture of the air, a tornado can last a few minutes or over an hour. However, cool winds (called rear flank downdrafts) eventually wrap around the tornado and cut off the supply of warm air that feeds it. The tornado thins out into the "rope-like" stage and dissipates a few minutes later.

Most tornadoes have wind speeds of less than 177 kph (110 mph), and are about 76 meters (250 feet) across. They can travel for several kilometers before dissipating. However, the most powerful tornadoes can have wind speeds of more than 482 kph (300 mph) and be more than 3 kilometers (2 miles) across. These tornadoes can travel across the ground for dozens of kilometers and through several states. These violent storms occur around the world, but the United States is a major hotspot with about a thousand tornadoes every year. "Tornado Alley," a region that includes eastern South Dakota, southern Minnesota, Nebraska, Kansas, Oklahoma, northern Texas, and eastern Colorado, is home to the most powerful and destructive of these storms. The most extreme tornado ever recorded occurred on March 18, 1925. This "Tri-State Tornado" sped for 338 kilometers (219 miles) through Missouri, Illinois, and Indiana. The tornado destroyed local communications, making warnings for the next town nearly impossible. The Tri-State Tornado killed 695 people in 3.5 hours. The best protection against a tornado is early warning. In areas where tornadoes are common, many communities have tornado warning systems. In Minnesota, for example, tall towers throughout neighborhoods sound an alarm if a tornado is nearby.

Measuring Winds

Wind is often measured in terms of wind shear. Wind shear is a difference in wind speed and direction over a set distance in the atmosphere. Wind shear is measured both horizontally and vertically. Wind shear is measured in meters per second times kilometers of height. Under normal conditions, the winds move much faster higher in the atmosphere, creating high wind shear in high altitudes.Engineers must consider an area's average wind shear when constructing buildings. Wind shear is higher near the coast, for example. Skyscrapers must account for this increased wind by having a stronger foundation or being engineered to safely "sway" with the wind. The amount of force that wind is generating is measured according to the Beaufort scale. The scale is named for Sir Francis Beaufort, who established a system for describing wind force in 1805 for the British Royal Navy. The Beaufort scale has 17 levels of wind force. "0" describes conditions that are so calm that smoke rises vertically. "12" describes a hurricane, and "13-17" are reserved only for tropical typhoons, the most powerful and potentially destructive wind systems.

An anemometer is a device for measuring wind speed. Anemometers are used with tornado data collectors, which measure the velocity, precipitation, and pressure of tornadoes. Tornadoes' strength is measured according to the Fujita scale. The scale has six categories that designate increasing damage. After the tornado has passed, meteorologists and engineers determine the tornado's strength based on its wind speed, width, and damage to vegetation and human-built structures. In 2007, the Enhanced Fujita Scale was established in the U.S.; it provides more specific effects of the tornado to determine its destructive power. The Enhanced Fujita Scale has 28 categories, with the strongest cataloging damage to hardwood and softwood trees. Hurricanes are measured using the Saffir-Simpson scale. In addition to tropical depressions and tropical storms, there are five categories of hurricanes. The most powerful, Category 5, is measured by winds whipping at 252 kph (157 mph). Tropical cyclones and typhoons are often measured using other scales, such as Japan's Tropical Cyclone Intensity Scale, which measures a typhoon as winds at 118 kph (73 mph).

Impact on Climate

Wind is a major factor in determining weather and climate. Wind carries heat, moisture, pollutants, and pollen to new areas. Many daily weather patterns depend on wind. A coastal region, for instance, undergoes changes in wind direction daily. The sun heats the land more quickly than the water. Warm air above the land rises, and cooler air above the water moves in over the land, creating an inland breeze. Coastal communities are usually much cooler than their inland neighbours. San Francisco is a coastal city in "sunny California," and yet the author Mark Twain noticed that "the coldest winter I ever spent was a summer in San Francisco!" Wind affects the climate of a mountainous area differently. Rain shadows are created as wind interacts with a mountain range.

As wind approaches a mountain, it brings moisture with it, which condenses as rain and other precipitation before coming over the crest of the mountain. On the other side of the mountain, dry "downslope winds" can speed through mountain passes at nearly 160 kph (100 mph). One of the most familiar of these downslope winds is the Föhn. Föhn winds—nicknamed "snow-eaters"—develop as air descends over the Alps, creating a warmer climate in central Europe. Winds also help drive ocean surface currents around the world. The Antarctic Circumpolar Current transports cold, nutrient-rich water around Antarctica. The Gulf Stream brings warm water from the Gulf of Mexico up the East Coast of North America and across the Atlantic to Northern Europe. Due to the Gulf Stream, Northern Europe enjoys a much warmer, milder climate than other areas at similar latitudes, such as the U.S. state of Alaska.

Impact on Ecology

Wind has the power to move particles of earth—usually dust or sand—in great quantities, and over far distances. Dust from the Sahara crosses the Atlantic to create hazy sunsets in the Caribbean. Winds transport volcanic ash and debris for thousands of kilometers. Winds carried ash from the 2010 eruption of Eyjafjallajökull, a volcano in Iceland, as far west as Greenland and as far east as Great Britain. The massive 1883 eruption of Krakatoa, an island volcano in Indonesia, had even more dramatic atmospheric results. Winds carried volcanic ash and debris high in the atmosphere across the globe. Europe endured years of cold, damp summers and pink sunsets. Wind's ability to move earth can erode the landscape. In some cases, this takes places in the desert, as sand dunes migrate and change shape over time. The wind can also pick up massive amounts of sand and "sandblast" rock formations into stunning sculptures. The Altiplano region of South America has dramatically shaped ventifacts—rocks carved by the wind-driven sand and ice. The wind's power to erode the land can be detrimental to agriculture. Loess, a sediment that can develop into one of the richest soils for farming, is easily swept up by wind. Even when farmers take precautions to protect it, the wind can erode up to 2.5 kilograms of loess per square meter (1.6 pound per square foot) every year. The most famous example of this devastating windstorm is probably the Dust Bowl of 1930s North America. Dust Bowl storms could reduce visibility to a few feet, and earned names like "Black Blizzards." Millions of farmers, especially those in the U.S. states of Oklahoma, Arkansas, and Texas, lost their land when they were unable to harvest any crops.

However, devastating to the economy, wind is an important way plants disperse seeds. This form of seed dispersal is called anemochory. Plants that rely on anemochory produce hundreds and even thousands of seeds. Seeds are carried by the wind to distant or nearby places, increasing the spread of the plant's genetics. Some of the most familiar seeds dispersed by the wind are those of the fuzzy dandelion.

Wind Energy

Wind has been used as a source of energy for more than a thousand years—it has pushed ships around the globe and been captured in windmills to pump water; it has turned giant stones to grind grains, make paper, saw logs, and crush ore. Today, most wind energy is used to generate electricity for homes, businesses, hospitals, schools, and industry. Wind is a renewable resource that does not directly cause pollution. Wind energy is harnessed through powerful turbines. Wind turbines have a tall tubular tower with two or three propeller-like blades rotating at the top. When the wind turns the blades, the blades turn a generator and create electricity. Often, wind turbines are collected in windy areas in arrays known as wind farms. Many wind farms have been established on mountains, in valleys, and offshore, as the air from the ocean interacts with land-air. Some people think wind turbines are ugly and complain about the noise they make. The slowly rotating blades can also kill birds and bats—but not nearly as many as cars, power lines, and high-rise buildings. The economic drawback to wind farms, however, is the wind itself. If it's not blowing, there's no electricity generated. Still, use of wind energy has more than quadrupled between 2000 and 2006. Germany has the most installed wind energy capacity, followed by Spain, the United States, India, and Denmark. Development is also growing quickly in France and China. Industry experts predict that if this pace of growth continues, by 2050, one-third of the world's electricity needs could be met by wind.

13.1 WIND

Wind in simple terms is nothing but moving air. Air moves because of pressure difference between two places. Wind blows from high pressure area to low pressure area. Pressure difference is created due to uneven heating of the earth surface. The horizontal movement of air is called wind and vertical movement of air is called current. Wind helps in transfer of heat and moisture from one place to another. Winds are the means by which the atmosphere attempts to balance the uneven distribution of pressure over Earth's surface. Our global wind system transports energy poleward to help maintain an energy balance, otherwise the equatorial regions would get hotter and the polar regions colder through time. The wind at the surface experiences friction. In addition, rotation of the earth also affects the wind movement. The force exerted by the rotation of the earth is known as the Coriolis force. Thus, the horizontal winds near the earth surface respond to the combined effect of three forces – the pressure gradient force, the frictional force and the Coriolis force. In addition, the gravitational force acts downwards. Centripetal acceleration produces a circular pattern of flow around centers of high and low pressure.

Pressure Gradient Force

The atmospheric pressure in itself is not a force. The difference in atmospheric pressure produces pressure gradient force.

The rate of change of atmospheric pressure with respect to distance at same elevation is known as pressure gradient (Δp / d).

Wind speed is directly proportional to pressure gradient. It acts in a direction from higher pressure to lower pressure. Thus, wind will blow from high to low pressure area.

Mathematically it can be expressed as

$$PGF = (1 / \rho).(\Delta p / d)$$

Where PGF is pressure gradient force, Δp is difference of pressure between two places, d is horizontal distance, and ρ is density of air. If the isobars are close to each other, the pressure gradient will be more and thus wind speed will be strong. If the isobars are widely spaced, the gradient will be less and thus wind speed is mild.

Coriolis Force (CF)

Because of the rotation of the earth about its axis, there is additional force acting on moving air. This force is known as coriolis force. The main characteristics of coriolis force are:

- It acts only at the right angles to the direction of moving air.
- It deflects wind to the right in the northern hemisphere and left in southern hemisphere.
- It only affects the direction of air motion/ wind. This coriolis force increases with wind velocity, air mass and an increase in latitude.
- At given latitude the magnitude of coriolis force is directly proportional to the speed of motion.

Mathematically the coriolis force (CF) can be represented by

$$CF = 2\Omega \text{ V Sin } \Phi$$

Where Ω is angular velocity of earth, V is wind speed and Φ is latitude

At equator, Φ = 00, Sin Φ =0, hence CF= 0; at 300, Sin Φ =0, hence CF= 0.5 and at poles Φ = 900, Sin Φ =1, hence CF= 1.

Coriolis force can be is a maximum (100%) at the poles but decreases to zero at the equator.

Frictional Force

The irregularities of the earth's surface offer resistance to the wind movement in the form of friction. Whenever air moves relative to the earth's surface, frictional drag tends to slow the air and also alter its direction of movement. The frictional force is greater over rough surfaces such as forest or city but less over the water surface. The effect of friction is most pronounced near the surface. It becomes relatively insignificant at elevation greater than 1 km above the surface. Frictional force acting on air is viscous force which arises when a layer of air moves over another.

Centrifugal Force

When air particles moves in a circular path, an acceleration towards the centre of the rotation is resulted.

$$C = \frac{mV^2}{r} C = \frac{mV^2}{r}$$

Where m is mass, V is velocity and r is radius of curvature

Due to this acceleration, a centrifugal force operates to through it outward radially.

Depending on the nature of forces acting on the wind, there are three main types of winds.

- Geostrophic Wind
- Gradient Wind

Geostrophic Wind:

When the pressure gradient force balances the Coriolis force the wind blows along the isobars (parallel to isobars), and the resulting wind is known as 'Geostrophic wind' (Figure 1). Geostrophic wind occurs in higher altitudes in the atmosphere where the frictional effects are negligible or insignificant.

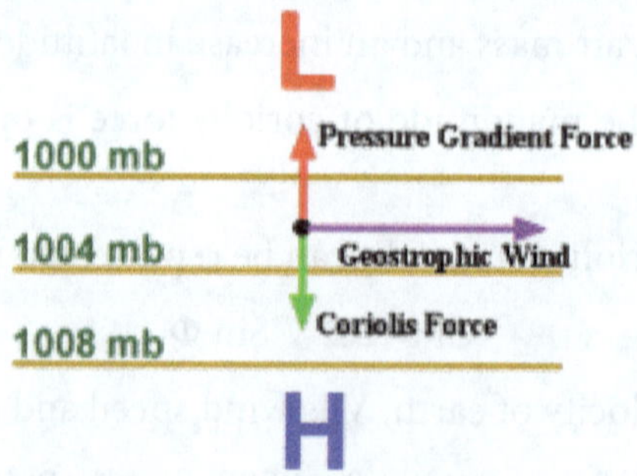

Figure 1: Geostrophic wind

Mathematically Geostrophic wind (Vg) can be expressed as a balance between CF = PGF Thus,

$$2\Omega \, Vg \, Sin \, \Phi = \frac{1}{\rho} * \frac{\Delta p}{d} 2\Omega \, Vg \, Sin \, \Phi = \frac{1}{\rho} * \frac{\Delta p}{d}$$

Therefore, geostrophic wind (Vg)

$$Vg = \frac{1}{\rho . 2\Omega Sin \, \Phi} * \frac{\Delta p}{d} \quad Vg = \frac{1}{\rho . 2\Omega Sin \, \Phi} * \frac{\Delta p}{d}$$

$$Vg = \frac{1}{\rho.f} * \frac{\Delta p}{d} Vg = \frac{1}{\rho.f} * \frac{\Delta p}{d}$$

Where f is coriolis parameter

$$(2\Omega \text{Sin } \Phi)(2\Omega \text{Sin } \Phi).$$

Gradient Wind:

When the wind blows along the curved path, the balance is between the pressure gradient force on one hand and the Coriolis force and the centrifugal force together on other hand. This is called as Gradient wind which blow along the isobars in the curved path but the speed vary with the pressure system (low or high). In a low-pressure system (cyclonic circulation) a balanced flow is maintained in a curved path by the excess of the pressure gradient force over the Coriolis force giving the net centrifugal acceleration outward. In this case of a low pressure system or trough, the gradient wind blows parallel to the isobars at a less than geostrophic (sub geostrophic) speed (Figure 2).

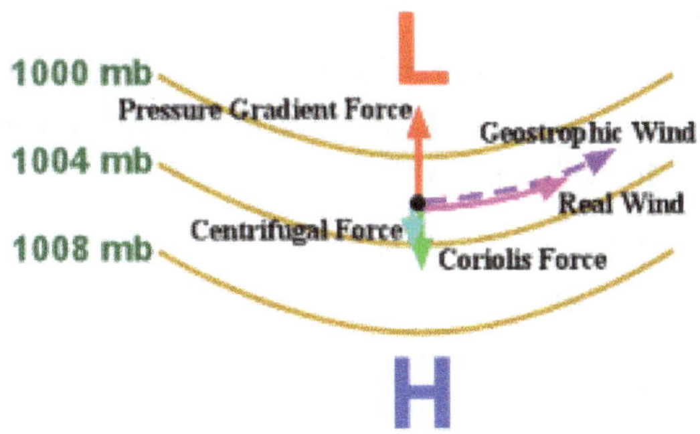

Figure 2: Gradient wind around a low-pressure system

Where as in a high-pressure (anti-cyclonic circulation) system it is the excess of Coriolis force over the pressure gradient that gives net centrifugal acceleration outward. This in turn increase the wind speed (Figure 4.3). This means that in a high-pressure system or ridge, the gradient wind blows parallel to the isobars faster than geostrophic (super geostrophic) speed.

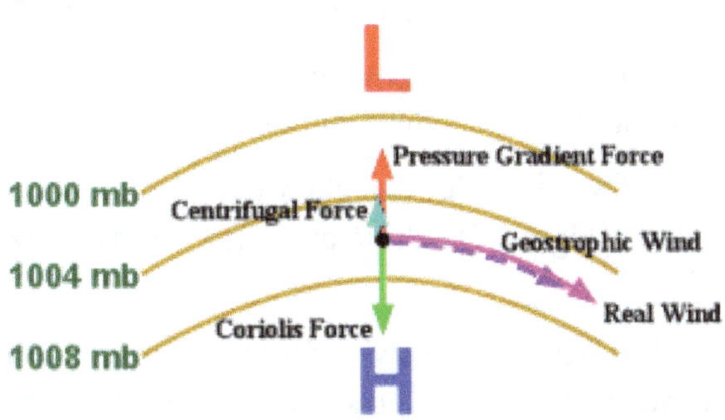

Figure.3: Gradient wind around a high-pressure system

Mathematically it can be expressed as

Pressure gradient = Coriolis force ± ± Centrifugal force

$$\frac{1}{\rho}\cdot\frac{\Delta p}{d} = 2\Omega\,V\sin\Phi + \frac{V^2}{r}\frac{1}{\rho}\cdot\frac{\Delta p}{d} = 2\Omega\,V\sin\Phi + \frac{V^2}{r}$$

Classification of Winds

Winds can be classified in three major groups

- Planetary winds or Primary wind or permanent wind or prevailing winds
- Secondary or periodic winds
- Tertiary or local winds

Planetary winds or Primary wind or permanent wind or prevailing winds

These are the winds which blow throughout the year over the globe. The pattern of the movement of the planetary winds is called the general circulation of the atmosphere. The pattern of planetary winds depends on the latitudinal variation of atmospheric heating; emergence of pressure belts; the migration of belts following apparent path of the sun; the distribution of continents and oceans, and the rotation of earth. The general circulation is comprised of three circulation cells (Figure 4).

Hadley Cell: Due to solar heating over the equator, the Inter Tropical Convergence Zone (ITCZ) is formed where air rises and reaches at the top of the troposphere and moves towards the poles. Due to cooling, the air subside at about 30° N and S, and create high pressure. On the land surface the air flows towards the equator and a complete circulation in formed. Such circulations from the surface upwards and vice-versa are over the tropics is called Hadley Cell.

Ferrel cell In the middle latitudes the circulation is that of sinking cold air that comes from the poles and the rising warm air that blows from the subtropical high. At the surface these winds are called westerlies and the cell is known as the Ferrel cell.

Polar cell At polar latitudes the cold dense air subsides near the poles and blows towards middle latitudes as the polar easterlies. This cell is called the polar cell.

These three cells set the pattern for the general circulation of the atmosphere. The transfer of heat energy from lower latitudes to higher latitudes maintains the general circulation.

There are three types of winds under Planetary winds or Primary winds or permanent winds or prevailing winds (Figure 5).. These are trade winds, prevailing westerlies and polar easterlies.

Trade winds The winds which are blowing from the sub-tropical high pressure areas (300 N or S) towards the equatorial low pressure belt are called trade winds. They flow as the North-east trades in the northern hemisphere and the South-east trades in the southern hemisphere. These are formed as a part of Hadley circulation cell.

Prevailing westerlies The westerlies are the winds blowing from the sub-tropical high pressure (300 N or S) belts towards the sub polar low pressure belts (600 N or S). They blow from south-west to north-east in the northern hemisphere and north-west to south-east in the southern hemisphere. These are called prevailing westerlies or simply westerlies.

Polar easterlies The Polar easterlies are dry, cold prevailing winds blowing from north-east to south-west direction in Northern Hemisphere and south-east to north-west in Southern Hemisphere. They blow from the polar high-pressure areas (900 N or S) to the sub-polar lows (600 N or S).

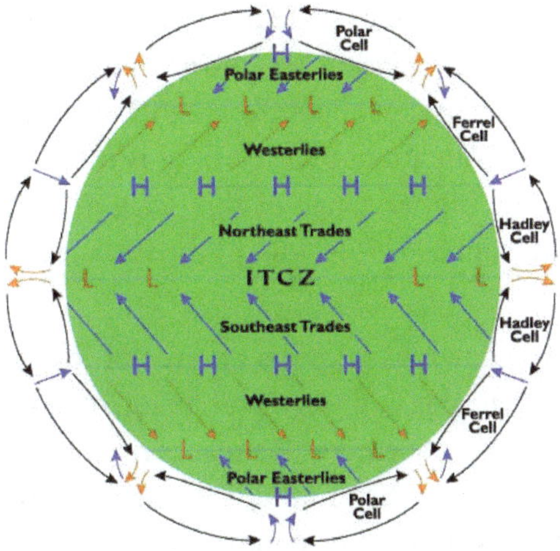

Figure 4: General circulation (Tri-cellular circulation) of the atmosphere

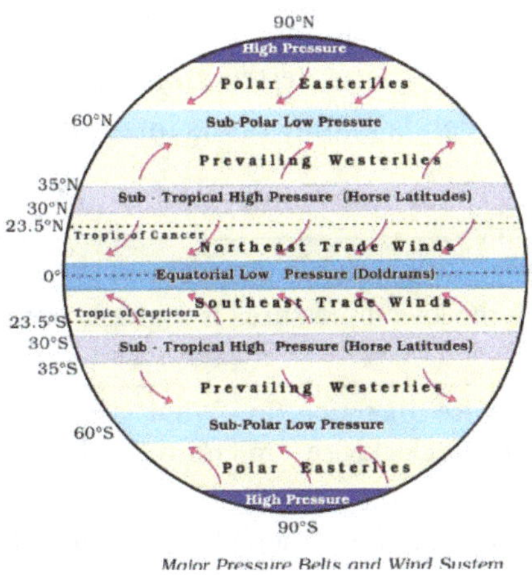

Figure 5: Primary wind system

Secondary or periodic winds

The secondary winds are those winds which blow due to seasonal change in temperature and pressure over a region. Monsoon, cyclones and anticyclones may be considered as secondary winds.

Monsoon: The monsoons are characterized by seasonal reversal of wind direction. During summer, the trade winds of southern hemisphere are pulled northwards by an apparent northward movement of the sun and by an intense low-pressure core in the north-west of the Indian subcontinent. While crossing the equator, these winds get deflected to their right under the effect of Coriolis force. These winds now approach the Asian landmass as south-west monsoons. Since they travel a long distance over a vast expanse of water, by the time they reach the south-western coast of India, they are over-saturated with moisture and cause heavy rainfall in India and neighbouring countries.

Cyclones and Anticyclones: Most of the day-to-day weather events are produced by traveling low or high-pressure systems called cyclones and anticyclones. These features are classified as secondary air circulation. The horizontal scale of these features is about 300- 3500 km. Their life time usually ranges from 2 to 3 days up to 10 to 12 days. Tropical cyclones whose size ranges from about 300- 600 km is smaller one of the secondary air circulations, where as in the higher range of this scale are extra tropical cyclones and anticyclones which extend from 800- 3500 km. At the surface convergence and divergence takes pace (Figure 6).

A cyclone, commonly known as a low, is an area of low pressure where air masses meet and rise. It typically indicates bad weather, like rain and clouds. Winds in a cyclone blow counter clockwise in the Northern Hemisphere and clockwise in the Southern Hemisphere. In a cyclone, air near the ground is pushed toward the low- pressure centre of the cyclone, and then rises upward, expanding and cooling as it moves. As it cools, the rising air becomes more humid, leading to cloudiness and high humidity within the cyclone. The main effects of tropical cyclones include heavy rain, strong wind, large storm surges near landfall, and tornadoes. The destruction from a tropical cyclone, such as a hurricane or tropical storm, depends mainly on its intensity, its size, and its location.

Table 1 : Difference between cyclones and anticyclones

SN	Cyclones	Anticyclones
1	It is low pressure area with the lowest pressure at the centre	It is high pressure area with the highest pressure at centre
2	Winds blow in anticlockwise direction in northern hemisphere and clockwise in southern hemisphere	Winds blow in clockwise direction in northern hemisphere and anticlockwise in southern hemisphere
3	It is associated with cloudy weather, heavy rain and high wind speed.	Little clouds with cool, dry air, clear weather are usually associated with anticyclone
6	Wind velocity increases from outer rim to the centre.	Wind velocities are much lesser than cyclones.
7	Horizontal size of the pressure system is around 300 to 600 km.	Horizontal size of the pressure system is extended from 800-1000 km.
8	At the surface air converges and moves upward	At the surface air divergence takes place

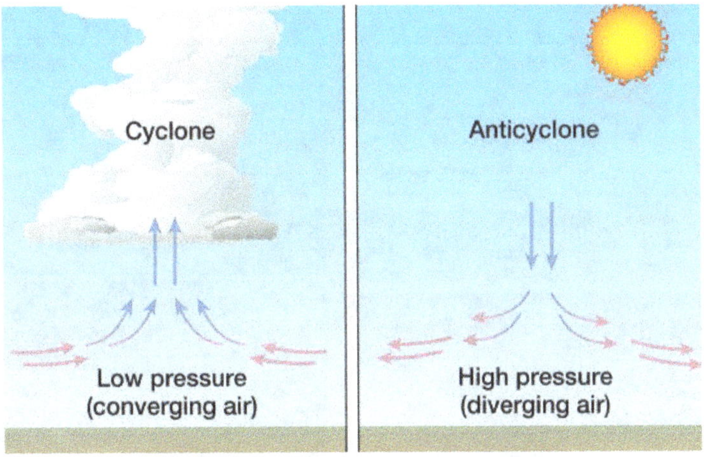

Figure 6: Cyclone and anticyclone

Tertiary or local winds

These are local air circulation patterns caused by local variations in the land topography and sea-land surface contrasts. Land and sea breezes, and mountain and valley breezes are some of the well-known local winds. The atmospheric motions which extend for horizontal distance from 1 to 2 km to 100 km are called the tertiary circulations or local winds. They are developed from local temperature differences and usually affect small areas. Following are the examples of local winds.

Land and sea breeze: The land and sea absorb and transfer heat differently. During the day the land heats up faster and becomes warmer than the sea. Therefore, over the land the air rises giving rise to a low-pressure area, whereas the sea is relatively cool and the pressure over sea is relatively high. Thus, pressure gradient from sea to land is created and the wind blows from the sea to the land as the sea breeze. In the night the reversal of condition takes place. The land loses heat faster and is cooler than the sea. The pressure gradient is from the land to the sea and hence land breeze results (Figure.7).

Table 2 : Difference between sea and land breeze

SN	Sea breeze	Land breeze
1	Occurs during day time	Occurs during nighttime.
2	Flows from sea to land.	Flows from land to sea.
3	Have more moisture	Do not have more moisture that sea breeze.
4	Stronger wind	Weaker than sea breeze.
5	Modifies weather on hot summer after noon.	Produces cooler winters and warmer summers.
6	It depends on topography of coastal region.	It depends on the topography of land.

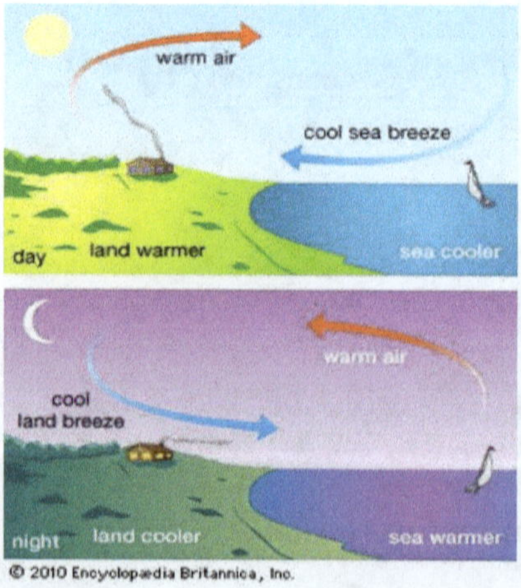

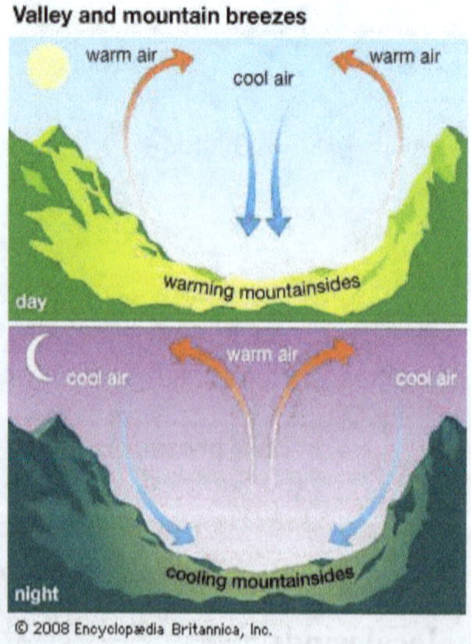

Figure 7: Land and sea breeze Figure 8: Mountain and valley breezes

Mountain and valley breezes: In mountainous regions, during the day the slopes get heated up and air moves upslope and to fill the resulting gap the air from the valley blows up the valley. This wind is known as the valley breeze. During the night the slopes get cooled and the dense air descends into the valley as the mountain wind. The cool air, of the high plateaus and ice fields draining into the valley is called katabatic wind (Figure.8.).

Another type of warm wind (katabatic wind) occurs on the leeward side of the mountain ranges. The moisture in these winds, while crossing the mountain ranges condense and precipitate. When it descends down the leeward side of the slope the dry air gets warmed up by adiabatic process. This dry air may melt the snow in a short time.

Table 3: Difference between mountain and valley breeze

SN	Mountain breezes	Valley breezes
1	Blows from mountain slopes to the valley base during night-time.	Blows from valley base to the up slope of mountain during daytime.
2	Cooling of air close to slope takes place.	Overheating of air adjustment to slope takes place.
3	It is known as "katabatic winds	It is known as "Anabatic winds".

14

OBSERVATION AND MEASUREMENT OF CLOUD

14.1 GENERAL INTRODUCTION

The observation of clouds and the estimation or measurement of the height of their bases above the Earth's surface are important for many purposes, especially for aviation and other operational applications of meteorology. This chapter describes the methods in widespread use.

14.1.1 Definitions Cloud

An aggregate of very small water droplets, ice crystals, or a mixture of both, with its base above the Earth's surface, which is perceivable from the observation location. The limiting liquid particle diameter is of the order of 200 μm; drops larger than this comprises drizzle or rain. With the exception of certain rare types (for example, nacreous and noctilucent) and the occasional occurrence of cirrus in the lower stratosphere, clouds are confined to the troposphere. They are formed mainly as the result of condensation of water vapour on condensation nuclei in the atmosphere. Cloud formation takes place in the vertical motion of air, in convection, in forced ascent over high ground, or in the large-scale vertical motion associated with depressions and fronts. Clouds may result, in suitable lapse-rate and moisture conditions, from low-level turbulence and from other minor causes. Human activity, such as aviation or industry, can also result in cloud formation, by adding condensation nuclei to the atmosphere. At temperatures below 0 °C, cloud particles frequently consist entirely of water droplets supercooled down to about –10 °C in the case of layer clouds and to about –25 °C in the case of convective clouds. At temperatures below these very approximate limits and above about –40 °C, many clouds are "mixed", with ice crystals predominating in the lower part of the temperature range.

14.1.2 Cloud Amount

The amount of sky estimated to be covered by a specified cloud type (partial cloud amount), or by all cloud types (total cloud amount). In either case, the estimate is made to the nearest okta (eighth) and is reported on a scale which is essentially one of the nearest eighths, except that figures 0 and 8 on the scale signify a completely clear and cloudy sky, respectively, with consequent adjustment to other figures near either end of the scale.

14.1.3 Cloud Base

The lowest zone in which the obscuration corresponding to a change from clear air or haze to water droplets or ice crystals causes a significant change in the profile of the backscatter extinction coefficient. In the air below the cloud, the particles causing obscuration show some spectral selectivity, while in the cloud itself, there is virtually no selectivity; the difference is due to the different droplet sizes involved. The height of the cloud base is defined as the height above ground level. For an aeronautical meteorological station, the ground (surface) level is defined as the official aerodrome elevation.

14.2 CLOUD TYPE (CLASSIFICATION)

Various methods of cloud classification are used, as follows:

- In WMO (1975), division is made into cloud genera with 10 basic characteristic forms, with further subdivision, as required, into:
- Cloud species (cloud shape and structure);
- Cloud varieties (cloud arrangement and transparency);
- Supplementary features and accessory clouds (for example, incus, mamma, virga, praecipitatio, arcus, tuba, pileus, velum and pannus);
- Growth of a new cloud genus from a mother-cloud, indicated by the addition of "genitus" to the new cloud and mother-cloud genera – in that order, if a minor part of the mother-cloud is affected – and of "mutatus" if much or all of the mother-cloud is affected, for example, stratocumulus cumulogenitus, or stratus stratocumulomutatus;
- A classification is made in terms of the level – high, middle or low – at which the various cloud genera are usually encountered. In temperate regions, the approximate limits are: high, 6–12 km (20 000–40 000 ft); middle, surface–6 km (0–20 000 ft); and low, surface–1.5 km (0–5 000 ft). The high clouds are cirrus, cirrocumulus and cirrostratus; the middle clouds are altocumulus and altostratus (the latter often extending higher) and nimbostratus (usually extending both higher and lower); and the low clouds are stratocumulus, stratus, cumulus and cumulonimbus (the last two often also reaching middle and high levels);

For synoptic purposes, a nine-fold cloud classification is made in each of these three latter divisions of cloud genera, the corresponding codes being designated CH, CM and CL , respectively. The purpose is to report characteristic states of the sky rather than individual cloud types;

- Less formal classifications are made as follows:
  - In terms of the physical processes of cloud formation, notably into heap clouds and layer clouds (or "sheet clouds");
  - In terms of cloud composition, namely ice-crystal clouds, water-droplet clouds and mixed clouds.

14.2.1 Vertical Visibility

The maximum distance at which an observer can see and identify an object on the same vertical as him/herself, above or below. Vertical visibility can be calculated from the measured extinction profile, σ(h), as stated by WMO. The relationship, however, is less simple than for horizontal visibility, because σ may not be regarded as a constant value. Nevertheless, the $I\ (h = VV)/I_{I0} = 5\%$ rule can be applied. Taking into account this assumption, the vertical visibility can be expressed in a relation with σ(h), in which VV is represented intrinsically, i.e.:

$$\int_{h=0}^{h=VV} \sigma(h)\,dh = -\ln(5\%) \approx 3 \quad (1)$$

14.2.2 Units and Scales

The unit of measurement of cloud height is the metre or, for some aeronautical applications, the foot. The unit of cloud amount is the okta, which is an eighth of the sky dome covered by cloud.

14.2.3 Meteorological Requirements

For meteorological purposes, observations are required for cloud amount, cloud type and height of cloud base. For synoptic observations, specific coding requirements are stated in WMO (2011), which is designed to give an optimum description of the cloud conditions from the surface to high levels. From space, observations are made of cloud amount and temperature (from which the height of the cloud top is inferred). Measurements from space can also be used to follow cloud and weather development.

14.2.4 Observation and Measurement Methods

14.2.4.1 Cloud Amount

Traditionally, measurements of cloud amount were made by visual observation. Instrumental methods are now widely accepted and are used operationally in many applications for determination of cloud amount and height. The cloud amount in each identified layer and the total cloud amount in view of the observation point are determined. The total cloud amount, or total cloud cover, is the fraction of the celestial dome covered by all clouds visible. The assessment of the total amount of cloud, therefore, consists in estimating how much of the total apparent area of the sky is covered with clouds. The partial cloud amount is the amount of sky covered by each type or layer of clouds as if it were the only cloud type in the sky. The sum of the partial cloud amounts may exceed both the total cloud amount and eight oktas. The scale for recording the amount of cloud is that given in Code table 2700 in WMO (2011), which is reproduced below:

Code figure	Meaning	
0	0 okta or less, but not zero	0
1	1 oktas	1/10 or less, but not zero
2	2 oktas	2/10-3/10
3	3 oktas	3/10
4	4 oktas	4/10
5	5 oktas	5/10
6	6 oktas	6/10 710-8/10
7	7 oktas or more, but not 8 oktas	9/10 or more, but not 10/10
8	8 oktas	10/10
9	Sky oscured by fog and/ or other meteorological phenomena	
/	Cloud cover is indiscernible for reasons other than fog or other meteorological phenomena, or observation is not made	

14.2.4.2 Cloud-Base Height

The height of the cloud base lends itself to instrumental measurement, which is now widely used at places where cloud height is operationally important. However, the estimation of cloud-base height by human observer is still widespread. Several types of instruments are in routine operational use, as described in this chapter. An international comparison of several types of instruments was conducted by WMO in 1986, and is reported in WMO (1988). The report contains a useful account of the accuracy of the measurements and the performance of the instruments.

14.2.4.3 Cloud Type

At present, the only method for observing most cloud types is visual. Pictorial guides and coding information are available from many sources, such as WMO (1975, 1987), as well as from publications of National Meteorological Services.

14.3 ESTIMATION AND OBSERVATION OF CLOUD AMOUNT, CLOUD-BASE HEIGHT AND CLOUD TYPE BY HUMAN OBSERVER

14.3.1 Making Effective Estimations

The site used when estimating cloud variables should be one which commands the widest possible view of the sky, and it should not be affected by fixed lighting which would interfere with observations at night. In making observations at night, it is very important that the observer should allow sufficient time for the eyes to adjust to the darkness. There are, of course, occasions when it is very difficult to estimate cloud amount, especially at night. The previous observation of cloud development and general knowledge of cloud structure will help the observer to achieve the best possible result. Access to reports from aircraft, if available, can also be of assistance.

14.3.2Estimation of Cloud Amount

The observer should give equal emphasis to the areas overhead and those at the lower angular elevations. On occasions when the clouds are very irregularly distributed, it is useful to consider the sky in separate quadrants divided by diameters at right angles to each other. The sum of the estimates for each quadrant is then taken as the total for the whole sky. Code figure 9 is reported when the sky is invisible owing to fog, falling snow, etc. or when the observer cannot estimate cloud amount owing to darkness or extraneous lighting. During moonless nights, it should usually be possible to estimate the total amount by reference to the proportion of the sky in which the stars are dimmed or completely hidden by clouds, although haze alone may blot out stars near the horizon. The observer must also estimate the partial cloud amount. There are times, for example, when a higher layer of cloud is partially obscured by lower clouds. In these cases, an estimate of the extent of the upper cloud can be made with comparative assurance in daylight by watching the sky for a short time. Movement of the lower cloud relative to the higher cloud should reveal whether the higher layer is completely covering the sky or has breaks in it. It should be noted that the estimation of the amount of each different type of cloud is made independently of the estimate of total cloud amount. The sum of separate estimates of partial cloud amounts often exceeds both the total cloud amount, as well as eight eighths.

14.3.4 Estimation of Cloud-Base Height

At stations not provided with measuring equipment, the values of cloud-base height can only be estimated. In mountainous areas, the height of any cloud base which is lower than the tops of the hills of the mountains around the station can be estimated by comparison with the heights of well-marked topographical features as given in a contour map of the district. It is useful to have, for permanent display, a diagram detailing the heights and bearings of hills and the landmarks which might be useful in estimating cloud height.

Owing to perspective, the cloud may appear to be resting on distant hills, and the observer must not necessarily assume that this reflects the height of the cloud over the observation site. In all circumstances, the observer must use good judgment, taking into consideration the form and general appearance of the cloud.

The range of cloud-base heights above ground level which are applicable to various genera of clouds in temperate regions is given in the table below and refers to a station level of not more than 150 m (500 ft) above mean sea level. For observing sites at substantially greater heights, or for stations on mountains, the height of the base of the low cloud above the stations will often be less than indicated in the table below.

In other climatic zones, and especially under dry tropical conditions, cloud-base heights may depart substantially from the given ranges. The differences may introduce problems of cloud classification and increase the difficulty of estimating the height. For instance, when reports on tropical cumulus clouds of an obviously convective origin, with a base well above 2 400 m (8 000 ft) or even as high as 3 600 m (12 000 ft), have been confirmed by aircraft observations. It is noteworthy that, in such cases, surface observers frequently underestimate cloud heights to a very serious degree. These low estimates may be due to two factors, namely either the observer expects the cumulus cloud to be a "low cloud" with its base below 2 000 m (6 500 ft) and usually below 1 500 m (5 000 ft), or the atmospheric conditions and the form of the cloud combine to produce an optical illusion.

When a direct estimate of cloud-base height is made at night, success depends greatly on the correct identification of the form of the cloud. General meteorological knowledge and close observation of the weather are very important in judging whether a cloud base has remained substantially unchanged or has risen or fallen. A most difficult case, calling for great care and skill, occurs when a sheet of alto stratus covers the sky during the evening. Any gradual lowering of such a cloud sheet may be very difficult to detect, but, as it descends, the base is rarely quite uniform and small contrasts can often be discerned on all but the darkest nights.

Cloud genera	**Usual range of height of base**		**Wider range of height of base sometimes observed, and other remarks**	
Low	(m)	(ft)	(m)	(ft)
Stratus	Surface-600	Surface-2 000	Surface-1 200	Surface-4 000
Stratocumulus Cumulus	300-1 350	1 000-4 500	300-2 000	1 000-6 SOO
Cumulonimbus	300-1 500	1 000-5000	300-2 000	1 000-6 500
Middle	600-1 500	2 000-5 000	300-2 000	1 000-6 500
Nimbostratus		(km)		
		Surface-3	Surface-10 000	Nimbostratus is considered a middle cloud for synoptic purposes, although it can extend to other levels

Cloud genera	Usual range of height of base		Wider range of height of base sometimes observed, and other remarks
Altostratus Altocumulus	2-6	6 500-20 000	Altostratus may thicken with progressive lowering of the base to become nimbostratus
High Cirrus Cirrostratus Cirrocumulus	6-12	20 000-40 000	Cirrus from dissipating cumulonimbus may occur well below 6 km (20 000 ft) in winter Cirrostratus may develop into altostratus
Note: a For stations ove 150 m above sea level, the base of low-level clouds will often be less than indicated.			

14.3.5 Observation of Cloud Type

Observation of cloud type is still widely performed by human observers. Pictorial guides and coding information are available from many sources, such as WMO (1975, 1987), as well as from publications of National Meteorological Services.

14.4 INSTRUMENTAL MEASUREMENTS OF CLOUD AMOUNT

Multiple types of ground-based operational sensors are available to measure total cloud amount. Measurements from space-borne radiometers in the visible band, supplemented by infrared images, can be used to estimate cloud amounts over wide areas, even though difficulties are often experienced, for example, the inability to distinguish between low stratus and fog. Amounts of low cloud within the range of a ceilometer can be estimated by measuring the proportion of elapsed time occupied by well-identified layers and assuming that these time averaged results are representative of the spatial conditions around the observing site. For synoptic meteorology, this technique is satisfactory in many cases but for airfield observations it can lead to significant errors in the estimation of cloud amount over the airfield. For automatic weather stations in the United States, a "clustering" technique has been developed using data from ceilometers. Other countries, like Sweden and the Netherlands, have introduced similar techniques in their operational observations. Other instruments used to measure cloud amount are pyrometers which may sample in multiple fixed directions and/or scan the sky, and sky cameras that are designed specifically for this purpose.

14.4.1 Using a Laser Ceilometer

Several meteorological services use time series of cloud-base measurements from laser ceilometers to determine cloud amount. This method has some advantages compared to manual observations. Using a ceilometer gives more consistent results. Also, output can be generated more frequently and there are no problems during night-time. However, there are also some drawbacks and large deviations can occur in situations with high, thin cirrus clouds when the performance of the ceilometer is reduced; when a moist layer is reported as a cloud base by the ceilometer; when a ceilometer detects no cloud base or at the wrong height during precipitation; and when the ceilometer reports a cloud base at the lowest elevation during shallow fog. This method relies on the clouds to move over the field of view of the instrument. Clouds do not always move in that way. Even if clouds do move across the field of view of the ceilometer, these clouds may not be representative of the total sky. Thus, the time series of the cloud base may not always represent the total sky, on which the reporting of cloud cover should be based. Agreements (within 2 okta) between this method and manual observation of total cloud amounts are typically 80%–90%, as found for coastal stations at mid-latitudes. However, most differences can be attributed to the limited spatial representativeness of a ceilometer sampling only a small area directly overhead.

Some airports are equipped with several ceilometers and a multiple-ceilometer sky condition algorithm. However, evaluation at an airport has shown only small improvements when using three ceilometers compared to one. This indicates that monitoring three points instead of one is still not sufficient to get a representative value for the entire sky. In the United States National Weather Service's Automated Surface Observing System (ASOS), the cloud height indicator compiles samples of backscatter return signals every 30 s and determines the height of valid cloud "hits". Every minute, the last 30 min of 30 s data are processed to give double weighting to the last 10 min in order to be more responsive to recent changes in sky condition. The data are then sorted into height "bins". Each minute, if more than five height bin values have been recorded (during the last 30 min), the cloud heights are clustered into layers using a least-square statistical procedure until there are only five bins remaining (each bin may have many hits in it). These bins, or clusters, are then ordered from lowest to highest height. Following this clustering, the ASOS determines whether clusters can be combined and rounded, depending on height, into meteorologically significant height groups. The resulting bins now are called "layers" and the algorithm selects up to three of these layers to be reported in the METAR/SPECI in accordance with the national cloud layer reporting priority.

The amount of sky cover is determined by adding the total number of hits in each layer and computing the ratio of those hits to the total possible. If there is more than one layer, the hits in the first layer are added to the second (and third) to obtain overall coverage. For reporting purposes, the ASOS-measured cloud amount for each layer is then converted to a statistical function equivalent to a human observation.

The algorithm also tests for total sky obscuration based on criteria of low surface visibility and a high percentage of "unknown hits" at low levels.

A sky condition algorithm has also been developed for use where cloud formation (or advection) typically occurs in (or from) a known location and results in significant concurrent differences in sky conditions over an airport. This meteorological discontinuity algorithm uses input from two cloud-height indicator sensors. The primary sensor is sited near the touchdown zone of the primary instrument runway. The second sensor is typically sited 3 to 6 km (2 to 4 miles) from the primary sensor, upwind in the most likely direction of the advection, or closer to the fixed source of the unique sky condition. The second cloud-height indicator serves to detect operationally significant differences in sky conditions.

14.4.2 Using a Pyrometer

Pyrometers, or passive infrared radiometers, are basically remote-sensing infrared thermometers (8–14 μm). These can be used to observe elementary solid angles of the sky either by using multiple fixed sensors (for example, four fixed sensors used to sample the whole sky), by scanning the entire sky dome with a single sensor, or by a combination of the two methods (one manufacturer's design has 14 sensors across 180 degrees of elevation from one horizon to the opposite horizon, and a physical mechanism scans the azimuth). The downward thermal emission from the clouds and from the air column between clouds and the instrument is measured and the temperature of each sampled solid angle is derived from a combination of the Planck and the Stefan-Boltzmann laws. The infrared temperature can then be used to provide an indication of cloud presence in each sampled solid angle. The total proportion of sky containing cloud can then be derived and reported as the cloud cover.

Scanning pyrometers avoid the problems of representativeness of the measurement that is present in other methods, depending on the number of points sampled. Also, nocturnal observations are possible. A disadvantage is that fractioned and/or transparent "pixels" are difficult to classify.

14.4.3 Using a Sky Camera

Cameras specifically designed to measure cloud amount exist. They view the total sky using, for example, curved mirrors. The image from the sky is analysed by an algorithm that determines whether a cloud is present in each pixel using the measured colour. The sum of all pixels results in cloud amount. This method avoids the problems of representativeness of the measurement that can be present in some other methods. Some cameras use daylight and are thus not applicable at night. Cameras measuring in the infrared do not have this disadvantage, but these have a smaller field-of-view and are more expensive. Sky cameras require frequent maintenance in the form of cleaning of the optical surfaces.

14.5 INSTRUMENTAL MEASUREMENT OF CLOUD-BASE HEIGHT

Several methods exist for measuring cloud-base height. They are: using a laser ceilometer, using a rotating beam ceilometer, using a searchlight and using a balloon. The method currently most used is the laser ceilometer. This technique has great advantages over other technologies and should therefore be considered as the most appropriate.

14.5.1 Using a Laser Ceilometer

14.5.1.1 Measurement method

With the laser ceilometer, the height of the cloud base is determined by measuring the time taken for a pulse of coherent light to travel from a transmitter to the cloud base and to return to a receiver (principle: light detection and ranging, lidar). The output from a laser is directed vertically upwards to where, if there is cloud above the transmitter, the radiation is scattered by the hydrometeors forming the cloud. The major portion of the radiation is scattered upward but some is scattered downward and is focused in the receiver onto a photoelectric detector. The radiant flux backscattered to the receiver decreases with range according to an inverse-square law. The ceilometer (Figure 15.1) generally comprises two units, a transmitter-receiver assembly and a recording unit. The transmitter and receiver are mounted in a single housing, together with signal detection and processing electronics. The light source is generally a semiconductor laser with a wavelength in the near infrared. The optics of the transmitter are arranged to place the laser source and receiver detector at the focus of a conventional or Newtonian telescope system. The surfaces of the lens are given a suitable quarter-wavelength coating to reduce reflection and to provide high transmission of light. The transmitter aperture is sealed by a glass window that is anti-reflection, coated on its inner surface and angled so that rain will run off it. The receiver is of similar construction to the transmitter, except that the light source is replaced by a photodiode, and a narrowband optical filter is incorporated. The filter excludes most of the background diffuse solar radiation, thus improving the detection of the scattered laser radiation by day. The transmitter and receiver can be mounted side-by-side so that the transmitter beam and the receiver field of view begin to overlap at about 5 m above the assembly and are fully overlapped at a few hundred metres. Some systems use the same lens for the transmitted and received radiation, so that this problem is avoided.

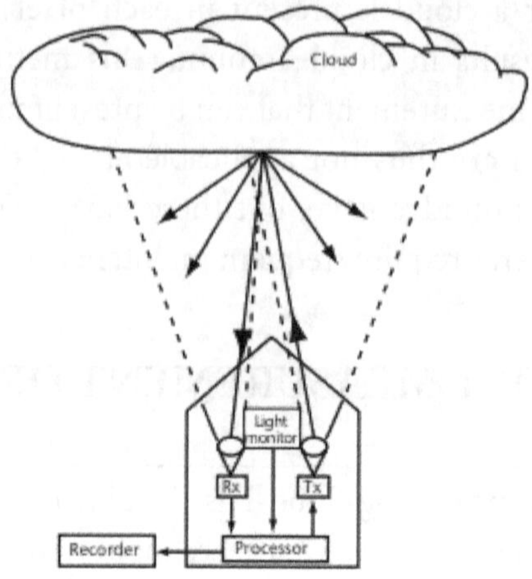

Figure 1. Typical laser ceilometer

The housing is provided with heaters to prevent condensation from forming on the optical surfaces, and the humidity within the housing can be reduced by the use of a desiccator. The top of the housing is fitted with a cover hood incorporating optical baffles that exclude direct sunlight. The output from the detector is separated into sequential "range gates", each range gate representing the minimum detectable height increment. A threshold is incorporated so that the probability of the instrument not "seeing" cloud, or "seeing" non-existent cloud, is remote.

14.5.1.2 Exposure and Installation

The unit should be mounted on a firm base, with a clear view overhead within a cone of approximately 30° about the vertical. If necessary, a rooftop site can be used with suitable adjustment of reported heights to ground level. Although laser ceilometers in operational use are designed to be "eye safe", care should be taken to prevent the casual observer from looking directly into the transmitted beam. To reduce the impact of strong reflecting raindrops, the beam with the telescope can be aligned about 5° from the vertical.

14.5.1.3 Sources of Error

There are four main sources of error as follows:

a. Ranging errors: These can occur if the main timing oscillator circuits develop faults, but, in normal operation, errors due to this source can be ignored;
b. Verticality of the transmitted/received beams: Provided that the instrument is aligned with the beam at better than 5° from the vertical, errors from this source can be ignored;
c. Errors due to the signal-processing system: Because a cloud base is generally diffuse and varies greatly in time and distance, complex algorithms have been developed to estimate a representative cloud-base height from the returned cloud signal. In conditions of fog (with or without cloud above) and during precipitation, serious errors can be generated. Thus, it is important to have an awareness of visibility and precipitation conditions to assess the value of ceilometer information. In conditions of well-defined stratiform cloud (for example, low stratocumulus), measurement errors are controlled solely by the cloud threshold algorithms and can be assumed to be consistent for a particular make of ceilometer;
d. Measurement range: Due to the limited power available from the laser, reflected radiation from high altitudes may have such low intensity that it cannot be detected. Therefore, cloud-base height from cirrus clouds may not always be observed.

In operational use and conditions of uniform cloud base, laser ceilometer measurements can be compared routinely with pilot balloon ascents, aircraft measurements and, at night, with cloud searchlight measurements. Intercomparisons of laser ceilometers of different manufacturers have been carried out extensively. During the WMO International Ceilometer Intercomparison (WMO, 1988), for example, several designs of ceilometer were intercompared and comparisons made with rotating-beam ceilometers and pilot-balloon observations. The international intercomparison revealed that, using current technology, laser ceilometers provided the most accurate, reliable and efficient means of measuring cloud-base height from the ground when compared with alternative equipment.

14.5.1.4 Calibration and Maintenance

Most laser ceilometers are provided with a built-in capability to monitor transmitted output power and guard against serious timing errors. Calibration checks are normally confined to checking both the master oscillator frequency and stability, using external high-quality frequency standards and the output power of the transmitter. Calibration may also be performed by intercomparison (WMO, 1988). Pointing the ceilometer to a target at a known distance (for example, a tower) can be used to confirm the distance measurement of the instrument. Routine maintenance consists typically of cleaning the exposed optics and external covers and of replacing air filters when cooling blowers are provided.

14.5.2 Using a Rotating Beam

14.5.2.1 Measurement method

The rotating-beam ceilometer (RBC) involves the measurement of the angle of elevation of a light beam scanning in the vertical plane, at the instant at which a proportion of the light scattered by the base of the cloud is received by a photoelectric cell directed vertically upwards at a known distance from the light source (see Figure 2). The equipment comprises a transmitter, a receiver and a recording unit. The transmitter emits a narrow light beam of a 2° divergence, with most of the emitted radiation on the near infrared wavelengths, that is, from 1 to 3 μm. Thus, the wavelength used is small in comparison with the size of the water droplets in clouds. The light beam is swept in a vertical arc extending typically from 8° to 85° and is modulated at approximately 1 kHz so that, through the use of phase-sensitive detection methods, the signal-to-noise ratio in the receiver is improved. The receiving unit comprises a photoelectric cell and an angle-of-view restrictor; the restrictor ensures that only light that falls vertically downwards can reach the photoelectric cell. A pen in the recording unit, moving simultaneously with the transmitter beam, records when a cloud signal is received.

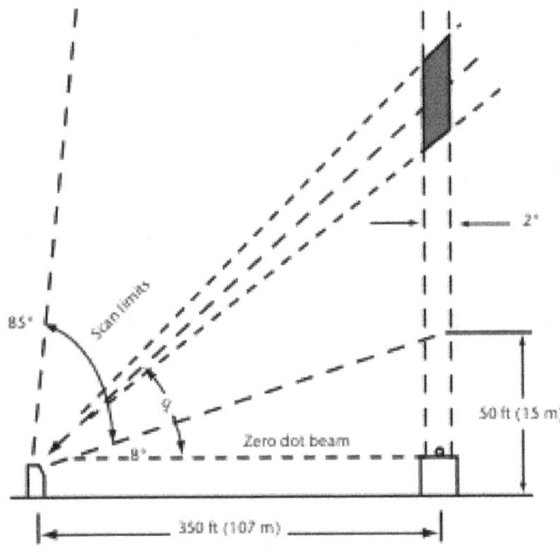

Figure.2. A Typical rotating-beam ceilometer

14.5.2.2 Exposure and installation

The transmitter and receiver should be sited on open, level ground separated by some 100 to 300 m and mounted on firm and stable plinths. It is extremely important that the transmitter scan in the same plane as the receiver. This is achieved by the accurate alignment of the optics and by checking the plane of the transmitter beam in suitable conditions at night.

14.5.2.3 Sources of error

Errors in the measurement of cloud-base height using an RBC may be due to the following:

- Beamwidth;
- Optical misalignment;
- Mechanical tolerances in moving parts;
- Receiver response.

Since in most designs the volume of intersection of the transmitter and receiver cone is very significant with a cloud height above 500 m, beamwidth-induced errors are generally the most serious. Some RBCs use a "threshold" technique to determine the presence of cloud, while others use a "peak" signal detection scheme. In either case, receiver sensitivity will affect reported cloud heights, giving rise to large errors in excess of stated operational requirements in some circumstances. These errors generally increase with indicated height. Rotating-beam ceilometers are very sensitive to the presence of precipitation. In moderate or heavy precipitation, the instrument can either indicate low cloud erroneously or fail to detect clouds at all. In foggy conditions, the light beam may be dissipated at a low level and the ceilometer can fail to give any useful indication of clouds, even when a low cloud sheet is present. Comparisons of RBCs and laser ceilometers have been carried out and widely reported. These have shown good agreement between the two types of ceilometers at indicated heights up to some 500 m, but the detection efficiency of the RBC in precipitation is markedly inferior.

14.5.2.4 Calibration and maintenance

The only maintenance normally undertaken by the user is that of cleaning the transmitter and receiver windows and changing the chart. The outside of the plastic windows of the transmitter and receiver should be cleaned at weekly intervals. A soft, dry cloth should be used and care should be taken not to scratch the window. If the transmitter lamp is replaced, the optical alignment must be checked. The transmitter and receiver levelling should be checked and adjusted, as necessary, at intervals of about one year.

14.5.3 Using a Searchlight

14.5.3.1 Measurement method

Using this method, illustrated in Figure 3, the angle of elevation, E, of a patch of light formed on the base of the cloud by a vertically-directed searchlight beam is measured by an alidade from a distant point. If L is the known horizontal distance in metres (feet) between the searchlight and the place of observation, the height, h, in metres (feet) of the cloud base above the point of observation is given as the following:

$$h\,L = \tan E \tag{2}$$

The optimum distance of separation between the searchlight and the place of observation is about 300 m (1 000 ft). If the distance is much greater than this, then the spot of light may be difficult to see; if it is much less, the accuracy of measuring a height above about 600 m (2 000 ft) suffers. A distance of 250–550 m (800–1 800 ft) is usually acceptable.

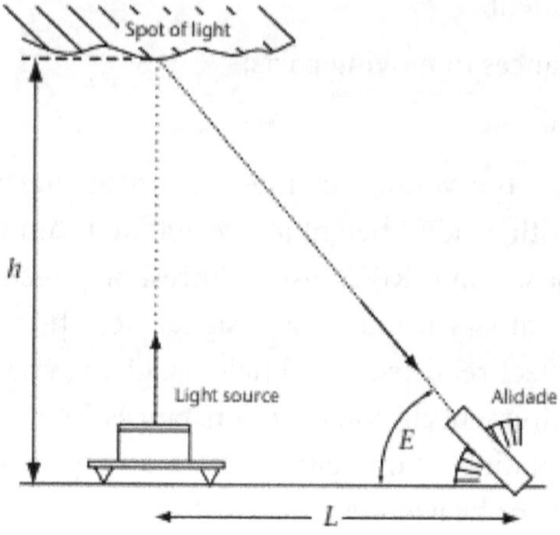

Figure 3. Principle of the cloud searchlight method

14.5.3.2 Exposure and installation

It is desirable to have a clear line of sight between the searchlight and the alidade, both of which should be mounted on firm, stable stands. Where there is a difference in the height above the ground between the searchlight and the alidade, a correction must be incorporated in the calculated heights. If a clear line of sight is not possible, any obstruction between the searchlight beam and the alidade should not be higher than 100 feet.

14.5.3.3 Sources of error

The largest source of error is due to uncertainty in the measured angle of elevation. Height errors due to small errors of verticality are insignificant. The absolute error Δh in the derived cloud height due to an error ΔE in the measured elevation is given by the following (L is assumed to be an accurately measured constant):

$$\Delta h = L \cdot \left(1/\cos^2 E \right) \cdot \Delta E = L \sec^2 E \cdot \Delta E \tag{3}$$

with E in radians ($1° = \pi/180$ rad). Note that Δh tends to infinity when $E \to 90°$. If $L = 1\,000$ ft (300 m) and $\Delta E = 1°$, the value of Δh is 17 ft (6 m) when $h = 1\,000$ ft (300 m), and Δh is about 450 ft (140 m) when $h = 5\,000$ ft (1 500 m). The relative error in h is given by:

$$\Delta h = L \cdot \left(1/\cos^2 E \right) \cdot \Delta E = L \sec^2 E \cdot \Delta E \tag{4}$$

with E in radians. $\Delta h/h$ is minimal when $E = 45°$ (or $h = L$).

14.5.3.4 Calibration and Maintenance

The focusing and verticality of the beam, should, if possible, be checked about once a month because the lamp filament is liable to undergo slight changes in shape with time. When a lamp is replaced, the adjustment for lamp position should be carried out since not all lamps are identical. The verticality of the beam should be checked during an overcast night with the aid of a theodolite. The check should be made from two positions, one near the alidade and the other at about the same distance away from the searchlight in a direction at right angles to the line joining the searchlight and the alidade (Figure 15.4). The azimuths of the searchlight and of the spot of light on the cloud should be measured as accurately as possible, together with the elevation of the spot of light. If the difference between the azimuth readings is A and the angle of elevation is E, the deviation ϕ of the beam from the vertical is given by:

$$\phi = \arctan\left(\tan A/\tan E \right) \approx A/\tan E$$
$$\left(\text{for } A \approx 1° \text{ or less} \right) \tag{5}$$

If the value of ϕ is more than 1° when viewed from the alidade, or more than 0.5° in the other position, these adjustments should be repeated until the necessary accuracy is obtained.

Focusing can be checked and adjusted on an overcast night by observing the diameter of the light spot on the highest cloud above the instrument. If necessary, the focus should be adjusted to minimize the spot diameter.

14.5.4 Using a Balloon

14.5.4.1 Measurement method

Cloud height may be measured in daylight by determining the time taken by a small rubber balloon, inflated with hydrogen or helium, to rise from ground level to the base of the cloud. The base of the cloud should be taken as the point at which the balloon appears to enter a misty layer before finally disappearing. The rate of ascent of the balloon is determined mainly by the free lift of the balloon and can be adjusted by controlling the amount of hydrogen or helium in the balloon. The time of travel between the release of the balloon and its entry into the cloud is measured by means of a stopwatch. If the rate of ascent is n metres per minute and the time of travel is t minutes, the height of the cloud above ground is not metres, but this rule must not be strictly followed. Eddies near the launch site may prevent the balloon from rising until sometime after it is released. Normally the stop-watch is started on the release of the balloon and, therefore, the elapsed time between when the balloon is released and the moment when it is observed to have left the eddies will need to be subtracted from the total time before determining the cloud height. Apart from eddy effects, the rate of ascent in the lowest 600 m (2 000 ft) or so is very variable. Although the height of the base of a cloud at middle altitude is sometimes obtained as a by-product of upper wind measurements taken by pilot balloons, the balloon method is mainly applicable to low clouds. Where no optical assistance is available in the form of binoculars, telescope or theodolite, the measurement should not be attempted if the cloud base is judged to be higher than about 900 m (3 000 ft), unless the wind is very light. In strong winds, the balloon may pass beyond the range of unaided vision before it enters the cloud. Precipitation reduces the rate of ascent of a balloon and measurements of cloud height taken by a pilot balloon should not be attempted in other than light precipitation. This method can be used at night by attaching an electric light to the balloon. For safety reasons, the use of candle lanterns is strongly discouraged.

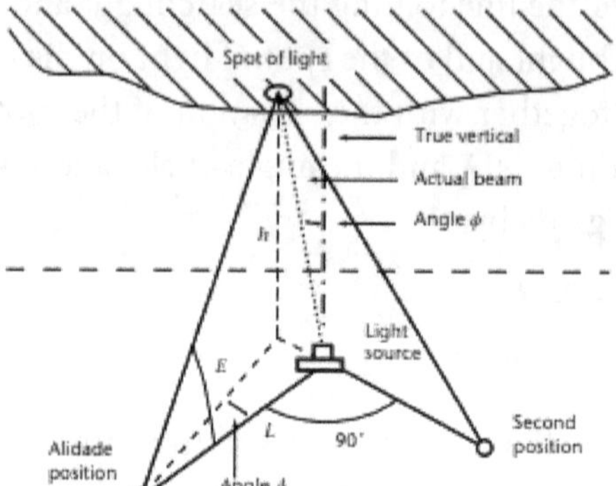

Figure.4. Checking the verticality of the searchlight beam

14.5.4.2 Sources of error

Measurements of cloud-base taken using a height balloon must be used with caution, since the mean rate of ascent of a balloon, especially in the first few hundred metres, may differ appreciably from the assumed rate of ascent (owing to the effects of vertical currents, the shape of the balloon, precipitation and turbulence).

14.6 INSTRUMENTAL MEASUREMENT OF CLOUD TYPE

Observation of cloud type is still generally performed by human observers. Only one automatic method to observe cloud type is known, which is specifically for detecting cumulonimbus/ towering cumulus. In this method, data from a precipitation radar and lightning detection network are used. The radar-reflectivity classes and the number of lightning discharges within a certain area are combined to give information on the presence of cumulonimbus and/or towering cumulus.

14.7 OTHER CLOUD-RELATED PROPERTIES

14.7.1 Vertical Visibility

Vertical visibility is defined as the maximum distance at which an observer can see and identify an object on the same vertical as him/herself. It can be calculated from the extinction profile of the atmosphere (WMO, 2010). Ceilometers may provide an estimate of vertical visibility, based on the integrated backscatter energy with range. WMO (1988) showed that this method frequently produces unreliable results. In practice, a vertical visibility report is often given by a ceilometer when the cloud-base requirements are not met, but when reflected light is received from a certain altitude.

15

MODIFICATIONS OF CROP MICROCLIMATE

15.1 INTRODUCTION

Microclimate is the climatic condition in a relatively small area, within a few meters or less above and below the surface and within canopies of vegetation. It is the climate very close to the ground surface.

- In case of crop plants, the microclimate extends from few tens of centimeters below the soil surface (root zone) to several tens of meters above the ground (crop canopy).
- In case of tree plants and forest areas, the microclimatic scales extend further deep in soil and hundreds of meters above the ground.
- This is the region where exchange of heat, moisture and mass takes place at the maximum rate. Soil and vegetation are the integral part of the microclimate.

Microclimatic modifications help to adjust the unfavourable conditions prevalent in the vicinity of the plants, allowing for greater crop growth and development, yield in the current situation of global warming and increased occurrence of extreme weather events. These Field level modifications of the physical environment, such as solar radiation, soil temperature, soil moisture, and wind speed, among other things, at the farm level have proven to be extremely beneficial for improved crop growth and yield performance. According to this study, the intensity and frequency of extreme weather elements are likely to increase in the future, and microclimatic modifications can be a very effective adaptation measure for managing extreme weather vulnerability and climatic risks in crop production to ensure food security and natural resource sustainability in the future.

- Microclimate refers to climatic conditions in a smaller area i.e. a few meters above or below the earth surface or within the crop canopy. It is the local climatic condition near the ground or area around the plants (up to about 2m height) resulting from the general climatic conditions.

- It also refers to the climate of small regions influenced by the effects of the relief, topography and the lower surface features, which create disparity between soil and air temperature, humidity and wind speed.
- Growth/development of plants mainly depends on their genetic constitution and environmental conditions.
- Climatic elements in the immediate vicinity of the plants are very important as they regulate and determine the physiological reactions. Lack of optimal climatic elements leads to disturbance in plant physio-logical and energy exchange processes resulting in undesired decrease in crop productivity.

15.2 DIVISION OF MICROCLIMATIC REGION

- The microclimatic region may be divided into **three layers** depending on the nature of transfer process takes place.
  - Surface layer
  - Skin layer 23 mm thickness above the ground surface maximum up to 1.5 meters height.
  - Surface boundary layer - 50100 m thickness above the skin layer.
- **Surface layer:** It is a plane separating two different media i.e. air (fluid) and soil (solid). It is an important site for energy and mass exchange and conversion. It receives energy from sun and acts as a source of heat energy for all the atmospheric processes.
- **Skin layer:** This is a thin layer adjacent to the surface ranging from few millimeters to 1.5 meters. The exchange of energy and mass is through the process of molecular conduction. Surface characteristic is dominant for exchange of energy. This is the layer, which has a direct significance to plants, animals and man. The environmental factors continually change in this layer. The nature of soil, type of plants grown, variable shading, and different wind protection methods are some of the factors responsible for these changes.
- **Surface boundary layer:** The layer is having the thickness of about 50 to 100 m. The exchange phenomena is through eddies or through turbulent transfer. Temperature variation during day and night are observed in this layer. The wind structure is governed by surface characteristics, in homogeneity, vertical distribution of temperature etc. The main transformations of energy i.e. from radiant energy to thermal energy and sensible heat to latent heat take place in the surface boundary layer. It is also the region in which the transformation of mass i.e. liquid water to gaseous water vapour occurs.
- All the crop management practices namely sowing time, planting method, row spacing, intercropping, tillage practices, mulch application, shelter belts and

irrigation management etc. affect the microclimate due to their effect on canopy temperature, wind speed, soil moisture, light interception and rate of waterloos etc. In other words, these microclimatic modifications affect the rate of exchange processes within the canopy as well as between the canopy and the surrounding air. Thus, by making some alterations / adjustments in crop management, we can modify the crop microclimate without any significant financial burden.

15.3 MICROCLIMATIC COMPONENTS

Microclimates are the dynamic, localised interactions between various processes in the surface layer, such as energy and matter exchange, radiation processes, and underlying surface effects.

15.3.1 Soil Moisture and Microclimate

One of the most important microclimate determinants is soil moisture. When soil moisture is present, the thermal conductivity and heat capacity of the soil are considerably boosted. As a result, soil moisture-rich locations have a more balanced microclimate with lower air and soil temperatures. This not only helps plants develop, but it also has an impact on weather and local rainfall patterns. There has been a lot of research in the last decade on the link between a lack of soil moisture and the incidence of severe temperatures and heat waves, both locally and regionally. Soil biotic life can thrive when there is enough moisture in the soil. Microorganisms help soil fertility by breaking down organic materials and releasing nutrients. When moisture occupies roughly 60% of the available soil moisture, optimal conditions are achieved. An abundance of water obstructs the delivery of oxygen, causing microbial activity to stall, cease, or turn anaerobic, negatively impacting plant growth.

15.3.2 Soil Characteristics and Microclimate

The relative amounts of clay, silt, and sand particles in the soil affect the texture. Clay particles are the tiniest, have the biggest surface area, and have the greatest ability to absorb water. Sand has the biggest particles and the least ability to absorb water. As a result, sandy soils have a lower moisture availability and a faster evaporation rate than clay soils. Clay soils, on the other hand, can harden in drought-prone locations, reducing infiltration and increasing runoff, reducing water availability. The soil structure is made up of soil texture, soil organic matter, and biological activity on the surface and below ground. The structure has to do with the development of micro- and macro-aggregates, which are the ways that distinct particles are held together. A good structure can reduce the wind and water erosion, as well as induce water infiltration and storage.

15.3.3 Soil Temperature and Microclimate

Incoming radiation, as well as the soil's thermal conductivity and heat capacity, determine the temperature of the soil. Soil colour has an impact on how much incoming radiation is absorbed or reflected. Darker soil absorbs a greater percentage of solar energy, whereas lighter soils reflect sunlight and are colder. During the day, heat transfer into the soil moves heat away from the direct surface, resulting in lower temperatures. When the surface temperature drops at night, the soil's heat transfer direction reverses, and heat is released to the surface, bringing the extremes back into balance. Over longer time scales, the same process happens, with heat being stored during warmer months and released during cooler months. The temperature of the soil promotes crop growth by supplying the warmth required by seeds, plant roots, and soil microorganisms. Plant growth can be hampered by high soil temperatures, while excessive temperatures can halt microorganism biological processes [9]. Low soil temperatures, on the other hand, impede plant water intake, hinder nitrification, diminishing soil fertility, and exacerbate desiccation when air temperatures are greater. High and low soil temperatures both affect plant evapotranspiration by increasing or lowering it.

15.3.4 Air Temperature and Microclimate

The most important factor of local air temperature is incoming and outgoing radiation. Local vegetation can promote transpiration, which lowers the temperature. Vegetation can also provide shade, preventing radiation from reaching lowerlying plants or surface levels (partially). Using the cooling effect of soil moisture to reduce total air temperature can result in higher crop yields by reducing extreme temperatures. The reflectivity of a surface determines how much sunlight is absorbed, which is referred to as albedo. It has a significant impact on determining local air temperatures, and it changes greatly depending on the weather. Local location has a significant impact on incoming radiation such as the direction in which a slope faces, influences the amount of energy received as well as shade. The albedo of the soil is determined by its moisture content. Because precipitation changes the local albedo and provides moisture for evaporation, the interaction between rainfall and air temperature is significant. A dry soil has a higher albedo than a wet soil in general. Croplands have a higher albedo than forests, which means they reflect more sunlight back into the sky and produce less surface heat [20]. Surface processes and qualities interact with temperature, moisture, and wind. Through shade, vegetation alters the radiation balance while also acting as a wind barrier. The upper crown has the highest air temperature, which happens one to two hours after local noon. Daytime temperatures are lower below the crown. The cooling of the earth's surface and the air near the ground, known as radiation cooling, causes minimum temperatures in the upper crown at night. This is especially true when the sky is clear, the wind is quiet, and the humidity is low.

Air humidity and microclimate High humid air absorbs water vapour more slowly than dry air, thus reduce plant transpiration. The existence of local wind is necessary to mix the environment since it moves damp air away from the vegetation. Dew formation was shown to be aided by mild breezes in unsheltered areas, but dew formation was found to be inhibited by moderate to high winds [33]. Dew production and duration are influenced by the presence of vegetation that acts as a windbreak or offers shade.

Windbreaks can help generate dew by reducing wind speeds, but they can also minimise it because local warmer air layers are not eliminated. Vegetation provides shade, which serves to lower local surface temperatures, increasing the odds of dew formation. The ability of wind to move air humidity can have a substantial impact on local humidity levels, both increasing and decreasing.

15.3.5 Wind and Microclimate

Wind has the ability to cool plants by removing the warm air-boundary layer that surrounds them. The removal of the layer and replacement with unsaturated air, causes higher transpiration, may also alter the plant's water intake. Furthermore, air circulation in the vegetation canopy is necessary for maintaining good CO_2 levels for growth, removing excess humidity, and lowering the general humidity level, minimising the risk of illness. Depending on the ambient temperature, wind can make temperatures warmer or cooler. In addition, many cereal crops are pollinated by the wind. Bacteria and fungi, like pollinators, rely on wind to move to new hosts, while insects use wind to expand their range . Dew formation is also influenced by local winds. Sediments carried by the wind collided with plant leaves and stems, causing structural damage. Wind have a cascading effect on the microclimate through the modification of vegetation potential and soil moisture storage capacity.

15.4MAJOR FIELD MICROCLIMATE MODIFICATIONS TECHNIQUES

Extreme weather conditions exist above and below ideal weather condition. Climate change is projected to amplify the frequency, intensity, and consequences of certain types of extreme weather events. These variables affect the development and growth of the plants. Rainfall/moisture, temperature, solar radiation, evaporation and evapotranspiration, and wind are all important meteorological characteristics. If one of those features is out of the box, the growth of the crop will suffer. Excessive rainfall, for example, causes floods, whereas a lack of rainfall causes drought. Cold wave conditions will occur if the temperature is significantly below normal. On the other hand, if the temperature is significantly higher than normal, a heat wave may occur. Similarly, cyclones have a negative impact on crop growth. Weather hazards are a significant threat to both crops and human activities. As a result, weather hazards must be modified using a variety of techniques in order to reduce losses.

15.4.1 Modification of Water Balance and Temperature Through Mulch Application

In the recent past, due to climate change, some areas have experienced severe drought, flooding, and other weather-related issues. On the other hand, scientists believe that relatively wet areas, such as the tropics and higher latitudes, will become wetter, while relatively dry areas, such as the subtropics (home to the majority of the world's deserts), will become drier. These vulnerabilities can be reduced by using mulch (organic or plastic). In a variety of ways, residue mulch left on or applied to the soil alters the crop microclimate above and below the soil surface, as well as within the mulch. Farmers all across the world use a variety of strategies, including residue retention, mulch application, and/or live mulch planting to modify solar radiation, reflection and absorption, shade, thermal radiation, temperature, humidity, wind/air movement, evaporation (crop and soil), soil moisture (surface and inside), and crop composition, structure, and growth. It may be beneficial to the seedbed because it prevents water loss from the soil surface, but heavy coatings may impede the germination of seeds that are sensitive to light and temperature variations. Depending on the properties of the residues, simultaneous changes in soil moisture and temperature, and changes in plant physiological activities as a result of the environmental circumstances provided by mulches. Mulch helps to store moisture, decrease the temperature of the soil, inhibit weed development, and promote fertility and soil health. Because of reduced soil evaporation and increased plant transpiration, mulching with crop leftovers boosted water-use efficiency by 20-30 %. Straw mulching has been observed to boost water use efficiency from 1.72 to 1.94 kg m3 in winter wheat. In drought situation using supplemental irrigation with mulch during the grain-filling stage resulted in a higher grain yield and can modify the microclimate. Mulch reduces the amount of solar energy that reaches the soil surface. During the winter, they raise the soil temperature, and during the summer, they lower it. During the summer, white and light-colored materials with high reflectivity are used to lower soil temperature, whereas dark-colored materials and black plastic mulches are used to raise soil temperature during the winter. Mulches reduce the rate of evaporation from the soil and conserve soil moisture for use by the plants by obstructing the exchange of water vapors. The benefits are greater during the summer / kharif season and on low-retentivity soils. Under mulch, potato had a much greater leaf area index, absorbed photosynthetically active radiation, yield, and water usage efficiency, which they believe is attributable to soil moisture conservation and a 4-6 o C reduction in soil temperature]. Mulch functions as a physical barrier to exchange processes on the soil's surface, altering the roughness of the soil/atmosphere boundary layer as well as the surface layer's dynamical heat and moisture properties. This straw layer is an excellent example of microclimate management and manipulation, both historically and more recently.

15.4.2 Modification Of Solar Interception Through Plant Density and Spatial Arrangement

The energy that powers the earth's climate system comes from the sun. Climate changes caused by variations in the composition and intensity of incident solar radiation hitting the Earth may be distinct from and complementary to those caused by man-made climate change on a global and regional scale. Solar variation appears to have a significant impact on regional climate in the current time. Row spacing has a significant impact on canopy temperature and crop radiation absorption. When row spacing is increased, the crop's ability to absorb radiation declines, more radiation falls on the soil surface, and the soil temperature rises. If row spacing is reduced, however, radiation interception by the crop increases and transmission to the soil surface decreases, lowering soil temperature. The cumulative effect of more effective tillers per unit area, grain number per spike, and grain weight amounted to a significantly higher grain yield in the N-S row direction. Lower canopy temperature and higher incidence of light in the upper canopy as the crop matured may have aided photosynthetic in the grain fill stage, resulting in increased grain yield. Planting the crop on N-E side of the ridge also reduce the solar load in crop plant by avoiding the direct sunrays.

15.4.3 Modification of Precipitation Through Artificial Rain Making

Short-term crop failures and long-term production let-downs are more likely as precipitation patterns change. Erratic rainfall with a high CV % result in erosion and water logging. Over the last 100 years, a trend of increasing monsoon seasonal rainfall has been observed along the west coast, norther Andhra Pradesh, and northwestern India (+10 to +12%), while a trend of decreasing monsoon seasonal rainfall has been observed over eastern, north-eastern India, and parts of Gujrat and Kerala (-6 to - 7%) . The most important requirement of a crop is moisture. Rainfall is critical in rainfed crops. The amount and distribution of rainfall throughout the crop's life cycle determines its growth. Moisture deficiency is harmful at any stage of the crop's life cycle, but it is especially lethal during the reproductive period. To mitigate the effects of a moisture deficit, artificial rain can be used. Because there may not be enough condensation nuclei in the atmosphere, artificial rain is based on the principle of introducing artificial condensation nuclei into clouds. This is referred to as weather modification. In 1930, Bergeron and Findeicen proposed a theory that rain drops form in clouds when a few ice crystals appear at temperatures below 0°C. After that cloud seeding is being popularised. Cloud seeding is a methodology for altering the weather. Artificial rain is created in this process by spraying dry ice or silver iodide aerosols into the upper part of the cloud in an attempt to stimulate precipitation and form rain. Aeroplanes and rockets can be used to stimulate the brain. Silver iodide is the most commonly used substance because it is inexpensive and readily available. The types of cloud seeding methods are as follows:

- Hygroscopic cloud seeding: In this method, the salts are dispersed in the lower part of the clouds using flares or explosives. When salt molecules come into contact with water, they grow in size.
- Static cloud seeding entails spraying a chemical such as silver iodide into the air or cloud. Silver iodide forms a crystal around it, allowing moisture to condense.
- Dynamic cloud seeding: This technique boosts the vertical air current, allowing more water to pass through the cloud's droplets, and they multiply quickly. They fall in the form of rain, hail, or snow from the sky.

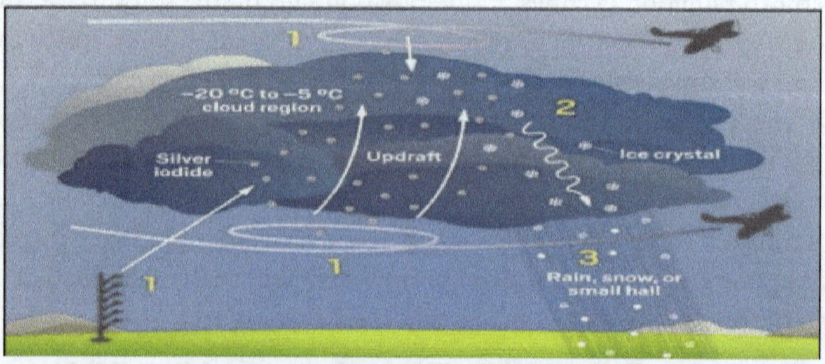

Modification of wind through windbreaks and shelterbelts "There is evidence for long-term changes in the large-scale atmospheric circulation, such as a poleward shift and strengthening of the westerly winds," according to the Intergovernmental Panel on Climate Change, and these changes are likely to continue. Windbreaks are a row or a group of trees, shrubs or structural elements used to block and direct the wind (e.g., fences). Windbreak trees protect a field from prevailing wind patterns, reducing wind speed significantly before it reaches the crops. By limiting wind speed, windbreaks alter the climate in the areas they cover. Reduced wind speed has a variety of consequences, including moderated soil and air temperatures, increased relative humidity, reduced evaporation and increased soil moisture, and changes in snow distribution. The temperature behind the shelterbelt is usually slightly higher because the wind's cooling influence is no longer there. Reduced wind exposure and increased humidity lower soil and crop evapotranspiration rates, allowing for more effective water use. Reduced kinetic impacts from wind, such as crop leaf damage and top soil erosion, is another good effect. As with other agroforestry practises, increasing water demand from windbreak trees is a significant problem. Windbreaks protect buildings from winter wind and summer sun, saving energy and lowering heating and cooling costs. Windbreaks can save up to 25% on winter heating expenditures. Windbreaks can save you a lot of money on air conditioning in the summer.

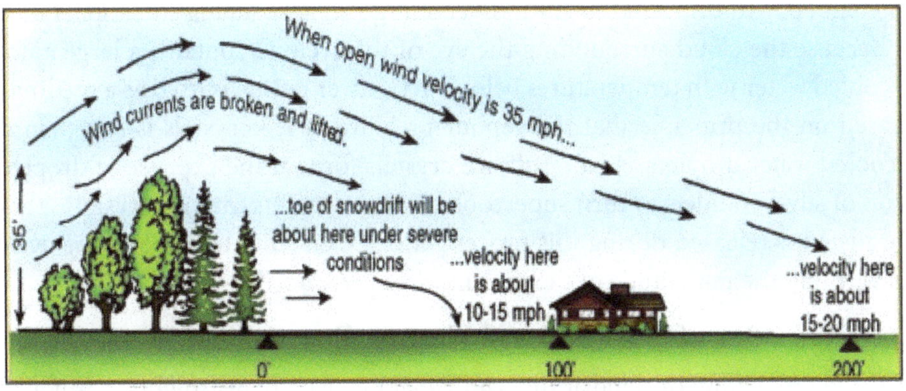

15.4.4 Modification of Plant Growth Through Fertilizer Management

Crop nutrients, whether organic or mineral, are the foods that nourish the plants, which in turn nourish the people. As a result, fertilisers are essential for food security. Damage due to frost is more occurred in unhealthy crops, and fertilisation improves plant health. Fertilization with nitrogen prior to a frost, on the other hand, promotes growth but increasing susceptibility to frost damage. To improve plant hardening, avoid using too much nitrogen fertiliser. Nitrogen fertilisation, on the other hand, gives the crop a new flush of growth after waterlogged situation. Phosphorus which is essential for cell division and tissue development be a boon after freezing injury. Phosphorus also helps in drought situation by improving the leaf water content, photosynthesis rate. Potassium has a positive effect on plant water regulation and photosynthesis. Some encouraging results have recently been obtained with postflowering foliar application of various nutrients on wheat yield. It was reported that spraying 0.5 % KNO_3 at 50% flowering stage of the crop resulted in higher grain and straw yield. Nowadays one of the reasons for crop yield reduction is a lack of micronutrients. Zinc is one of the micronutrients that plays an important role in most plants' metabolic activities. Zinc spraying in drought conditions increases the number of grains per spike, grain weight, and harvest index when compared to no spraying.

15.4.5Modification of Cyclone Through Cloud Seeding

Warmer sea surface temperatures (above 26.5° c) may increase the speed of tropical storm winds, causing more damage if they make landfall. Cyclones are one of the most dangerous weather hazards, wreaking havoc on agricultural crops in coastal areas. Cyclones have a negative impact on all human activities. These cyclones are also known as tropical cyclones, typhoons, or hurricanes. The main benefit of these cyclones is that they cause rainfall over land in dry areas, but excessive rainfall can cause flooding over a large area, especially near the coast. Because of the devastation caused by these weather systems, they must be modified. Cyclones can be modified by seeding the outer clouds surrounding the eye of the cyclone so that precipitation occurs before the mature stage is reached. An enormous amount of latent heat of condensation is released during precipitation. The latent heat has the tendency to disperse the storm over a large area, reducing the impact of violent force. Because the cloud surrounding the eye of the cyclone contains a large amount of super cooled water with temperatures below -4°C, silver iodide is used as a seeding agent. It is based on the principle that the vapour pressure of ice crystals is less than that of supercooled water droplets. As a result, ice crystals form at the expense of droplets. The addition of silver iodide can turn supercooled water droplets into ice crystals. The latent heat of fusion is released during this process. It has the ability to disperse the cyclone in such a way that the magnitude of the violent force is reduced.

15.4.6Modification of Frost Through Heaters, Fire and Smoke

Frost is the ice coating or deposit that can generate, usually overnight, in wet air under cold conditions. This happens when the Earth's surface temperature and any earth-bound object drops below zero degrees (freezing). For centuries, fire and smoke have been used in traditional agricultural systems. Heaters provide additional heat to compensate for energy losses. Heaters, in general, either raise the temperature of metal objects (e.g., stack heaters) or function as open fires. If enough heat is added to the crop volume to compensate for all energy losses, the temperature will not drop to dangerous levels. They are cost effective for high value crops (sugarcane, coffee, tea etc). Biomass residues (crop straw, residue, waste etc) are also used to heat the crop environment in windy condition. Straw burning in the wind creates a smoke layer over the crop surface that absorbs long-wave radiation released by the soil, protecting the mustard crop from frost by boosting the ambient temperature. Smoke particles are typically less than 1 μm in diameter, reflect visible light but are impervious to long wave radiation, and hence prevent rapid cooling of the ground surface [30]. In frost time even light watering or sprinkling helps in frost protection by releases latent heat and prevents the tissues of the plants it coats from freezing. Sprinkler irrigation provides good frost protection and raises the temperature of the microclimate. Sprinkler irrigation increases long-wave radiation and sensible heat transmission to the plants when compared to an unprotected crop. By wetting the soil, it becomes darkens, which increases the absorption of solar radiation and avoid the chilling injury. Using powerful blowers to reduce wind, preventing the formation of cold air accumulations, Crop covering or wrapping (high-value crops), Heat to keep the temperature from dropping too quickly, Smoke production to reduce radiation cooling are Some other Typical measures to prevent frost or reduce its severity.

15.4.7 Modification of Heat Waves Through Protected Cultivation

Heatwaves, or prolonged hot weather periods, can have significant social effects including higher deaths due to hot weather. Temperatures of days and nights are becoming more as climate change becomes more common. Increases in extreme heat events may result in an increase in heat-related illnesses and deaths, resulting in decreased productivity and lower-quality produce. Protected cultivation makes it possible to obtain increased crop productivity by maintaining a favorable environment for the plants by reducing the heat waves. The use of netting and other type of covering has been shown to restrict air movement around the growing seedlings in higher temperature [26]. Protected cultivation, on the other hand, is used to protect plants from harsh climatic conditions by providing optimal lighting, temperature, humidity, CO2, and air circulation for optimal plant growth and quality. Increased air temperatures inside the house combined with improved moisture status and root development, increased nutrient uptake, favouring leaf conductance and chlorophyll content.

Figure 5: Growing crops in greenhouse as protected cultivation (source: https://thumbs.dreamstime.com/z/strawberries-protectedcultivation)

Modification of flood through dam, and land-form change As a consequence of climate change, floods and droughts have become more frequent, posing a threat to traditional irrigation design and management systems. Construction of dams and detention basins can help to reduce flood peaks. Dams can hold a large amount of water during a flood and help to reduce the peak volume of water during a flood. Depending on the accommodating capacity of the river downstream of the dam, water stored in reservoirs created by dam construction can be allowed to flow down the stream under controlled conditions. These dams have aided in the reduction of flood peaks in downstream areas. Ponds, tanks, and surface storage structures, in addition to dams, help with flood control and water harvesting during dry seasons. Raised Beds are a revolutionary way to prevent water logging that increases productivity significantly. Permanent Raised beds and broad beds, are the land modification that helps to save 25- 35 % of irrigation water, improve water efficiency, and avoid excess water near the crop root zone.

15.5 MICROCLIMATIC MODIFICATIONS

- Microclimate modification includes an artificially induced changes in the composition, behaviour or dynamics of the atmosphere near the ground so as to improve the environment in which the crops are grown. By making such alterations, the microclimate can be made more favourable for better growth and yield of the crops.

- Microclimatic modification is any process in which all the adverse or stress conditions prevailing in the immediate vicinity of the plants are modified for the better crop growth, development and yield.
- Artificial control of plant environment to maintain the optimum conditions for better plant growth and crop production is future trend in agro meteorological research.
- Short-term adjustments at the farm level involve production techniques, such as changes in crop rotation and crop cultivars, changes in soil cultivation and tillage practices, a shift of sowing dates, adapted fertilization and crop protection measures.
- Long-term adaptations, on the contrary, include major structural changes of farm production systems and need careful agro-economic planning and realization at societal level; these adaptations also involve a set of sectors and stakeholders, such as policy, research, water and land planning.
- A change in the landscape structure, such as the introduction of windbreaks or hedges, influence the microclimate of crops in neighbouring fields, mainly by slowing down the wind speed.
- Major processes of microclimatic modifications include controlling of heat load, water balance, wind speed and modification of temperature and solar radiation.
- Heat load on crop canopy can be increased (during winter) called heat trapping, or decreased (during summer) called heat evasion.
- Heat evasion can be achieved by shading the crop, irrigation management or by using anti-transparent, whereas heat trapping can be achieved by adopting appropriate row direction or plant density.
- Controlling the water balance involves increasing the amount of water stored in the root zone (by strip cropping, contour cropping, ploughing, terracing or bunding), increasing infiltration (by increasing tillage or increasing row spacing), reducing soil evaporation (by using mulches and crop residues, plastic covers or thick cultivation crops), modifying transpiration (by using anti-transpirants, substances forming film on leaf surface, stomata closing materials and reflectants).
- Whereas, winds can be controlled by applying wind breaks or shelter belts. These protections reduce wind speed on the leeward side and save the plants from freezing injury and mechanical damage (Fig. 1).
- Mulching helps in regulating soil temperature and conserving soil moisture by preventing evaporation losses, hence saves the crop from harsh weather conditions.

- Row spacing and row orientation can be altered for efficient utilization of solar energy. Modified crop microclimate with improved irrigation management leads to increase heat and water use efficiency.
- Average daily temperature and relative humidity remain significantly higher in net house as compared to open conditions.
- Raised beds help in improving water use efficiency by saving irrigation water by 25-35%.
- Green house gas emissions reduce under no tillage treatment.
- Intercropping helps in improving PAR interception and RUE as compared to sole crop.
- Modification of solar radiation can be accomplished by increasing the surface absorptive power, reflective power of the surrounding objects, and exposure through site selection, increasing the radiant energy by fog dissipation and by adopting appropriate row direction.
- Similarly, soil temperature can be modified by mulch application and soil tillage etc. Mulch helps in reducing the soil temperature in summer and increasing during winter season. Similarly soil tillage by harrowing or inter-cultivation also helps in modifying the soil thermal regime.
- The protected cultivation under controlled environment like greenhouses (for temperature, humidity and CO_2 control), polyhouses (naturally ventilated or with controlled environment) and shade houses (for crop production in warm climate) provide other means for raising crops under controlled microclimatic conditions
- Different crops have specific climatic requirements at different growth stages. Thus, raising crops under favourable microclimate as per requirements of the crops, needs different management techniques so that sensitive growth stages of the crops can coincide with the favourable micro-environment.

15.6 FIELD CROP MICROCLIMATIC MODIFICATION

- Various agronomic manipulations and other microclimate modification options viz. alteration in sowing time, mulch application, row orientation and spacing, irrigation management, planting methods, tillage practices, appropriate cropping systems and installation of the net houses/screen houses etc. can be used to make the crop microclimate optimum for highest growth rate and productivity.
- **Sowing time:** Altering sowing time is an important production component that can be manipulated to counter the adverse effects of environmental stress. This is accomplished through shifting of sowing time so that any stress caused by environment is avoided during the critical stages of crop growth. Matching the

phenology of the crop to the duration of favourable environmental conditions by selecting the most appropriate sowing time to avoid the periods of stress is crucial for obtaining maximum yields under changing climate.

- **Mulching:** Mulch is a layer of material applied at the soil surface, which leads to conserve soil moisture, moderate soil thermal regime, reduce weed growth and improve fertility and soil health.
- Mulch obstructs the solar radiation reaching the soil surface. White- and light-coloured materials which increase the reflectivity, are used to decrease soil temperature during summer season, whereas, dark-coloured materials and black plastic mulches are used to increase soil temperature during winter season.
- Straw mulches can be used during both the seasons as they increase soil temperature during winter season and decrease during summer season. Due to obstruction of the exchange of water vapours, mulches decrease the rate of evaporation from the soil and conserve soil moisture for use by the plants.
- **Row spacing:** Row spacing has a great impact on canopy temperature and radiation interception by the crop. If row spacing is more, radiation interception by the crop decreases, more radiation falls on the soil surface and increases the soil temperature. On the other hand, if row spacing is less, radiation interception by the crop increased and its transmission towards soil surface decreases, which decreases the soil temperature.
- Appropriate row spacing and orientation help in trapping of radiation by the canopy, thus leading to higher radiation interception and crop productivity.
- **Irrigation management:** By applying irrigation soil moisture availability increases, which increases the evapotranspiration and decreases the soil as well as canopy temperature. Under very hot and dry conditions, light irrigation is a common practice to save crops from high temperature stress. Appropriate amount and time of irrigation results in managing heat and water stress and in improving yield and water use efficiency.
- **Method of planting:** Method of planting also affects the crop microclimate due to its effect on soil properties. Bed sowing produces significantly taller plants with greater biomass and grain yields in comparison to flat sowing.
- **Tillage** is the physical manipulation of soil, which is intended to destroy weeds, incorporate crop residues and amendments into soil, increase infiltration and reduce evaporation, prepare seedbed and break hard layer to facilitate root penetration.
- **Tillage practices** modify the crop microclimate by affecting the soil properties. Tillage increase the porosity and decreases the thermal conductivity of the soil and increases the soil temperature.

15.7 SHELTERBELTS / WINDBREAKS

- The main function of shelterbelts is to reduce the wind speed and cold advection. Due to reduction in wind speed, the evapotranspiration decreases and the temperature and moisture increases. Windbreaks modify the micro-climate by providing the shelter, which reduces the wind speed and turbulence. Wind breaks protect the plants from the high intensity winds. They modify the water table by reducing the water losses and hence they improve the crop productivity.
- Wind breaks also reduce the wind erosion. In dry farming, windbreak reduces crop evapotranspiration by modifying the aerodynamic component of the energy balance. There is a positive effect of windbreaks in mitigating evapotranspiration, hence their significant applicability under dry-farming systems.

15.8 CROPPING SYSTEMS

- Intercropping *i.e.* growing of two or more crops simultaneously on the same field such that the period of overlap is long enough to include the vegetative stage, has significant effect on microclimate and resource use efficiency.
- Intercropping, double cropping and other mixed cropping practices lead to enhanced efficiency of farm resources with sustainable crop production.
- Intercropping increases the radiation interception by the canopy and decreases its transmission towards the soil surface, thus also decreasing the soil temperature.
- Improved productivity thus results from either greater interception of solar radiation or higher light use efficiency or their combination.
- With Intercropping, microclimate within canopy can moderate temperature extremes, lower temperatures with reduced air movement leading to decreased evaporation rates and increased relative humidity, which is important in avoiding desiccation and providing favourable growth conditions even during the periods of moisture deficit.
- Although intercropping is also known to decrease the occurrence of weeds, insects and diseases, but modified microclimate especially moderate temperatures and increased relative humidity sometimes might become conducive for disease outbreak, especially fugal pathogens.

15.9 PROTECTED CULTIVATION

- Protected cultivation makes it possible to obtain increased crop productivity by maintaining favourable environment for the plants. The use of netting and other type of covering has been shown to restrict air movement around the growing seedlings in higher temperature.

- Elevated air temperatures inside the net along with improved moisture status and root development enhanced the uptake of nutrients such as potassium and nitrogen, thereby favouring leaf conductance and chlorophyll content.
- The net covering also offers a physical barrier reducing the occurrence of insect-pests. Higher relative humidity recorded under netting could also affect the feeding habit of sucking pests, hence lowering their population under netting.
- **Fogging system** could improve the climatic conditions under screen house and extend the growing season during adverse environmental conditions by moderating the extremes of microclimate during summer.

RELATED TERMINOLOGY

Advection: The horizontal transport of some property of the atmosphere or ocean, such as thermal energy, humidity, or salinity. In the context of meteorology, the related term convection generally refers to vertical transport.

Actinometer: A scientific instrument used to measure the heating power of radiation, particularly solar radiation.

Adiabatic process: Any idealized hypothetical process by which energy is transferred between a thermodynamic system and its surroundings only as work, without a corresponding transfer of heat or mass. In the context of meteorology, the assumption of adiabatic isolation is often used to explain the processes of adiabatic heating and adiabatic cooling, which involve changes in temperature as a parcel of air is compressed or expanded by interaction with its surroundings.

Aerobiology: The branch of biology that studies airborne organic particles, such as bacteria, viruses, fungal spores, pollen grains, and very small insects, which are passively transported by the air.

Aerography: The production of weather charts.

Aeronomy: The branch of meteorology that studies the upper regions of the Earth's or other planetary atmospheres, specifically their atmospheric motions, chemical compositions and properties, and interactions with the other parts of the atmosphere and with space.

Aerosol: A suspension of fine solid particles or liquid droplets in air or another gas. Examples of natural aerosols include mist, clouds, fog, and dust.

Air current: Any concentrated area of winds that develops due to differences in pressure and/or temperature between adjacent air parcels. They are generally divided into horizontal and vertical currents and exist at a variety of scales and in various layers of the atmosphere.

Air mass: A volume of air defined by its temperature and content of water vapor.

Air parce: lIn fluid dynamics, any amount of air that remains identifiable throughout its dynamic history while moving with an associated air flow.

Air-mass thunderstorm: Any thunderstorm that is generally weak and usually not severe. Such storms move relatively slowly, are short-lived, and often exist only as single cells (rather than in long continuous lines or complexes), but may still produce lightning and heavy rainfall. They derive their energy from solar radiation and commonly develop in temperate zones during summer afternoons.

Albedo: An index of the "reflectiveness" of a surface; a way of quantifying how much radiation is reflected back, as opposed to absorbed. Objects or surfaces with low albedo (closer to 0) absorb most of the radiation directed toward them, and those with high albedo (closer to 1) reflect most of it

Anemometer:A scientific instrument used to measure wind speed.

Anticyclone: Any large-scale air mass characterized by outward spiraling winds which circulate around a strong center of high atmospheric pressure. Surface-based anticyclones generally bring about cool, dry air and clear skies and are often implicated in weather phenomena such as fog and haze. Contrast cyclone.

Atlantic: hurricaneA tropical cyclone (locally known as a hurricane) that forms in the Atlantic Ocean and achieves one-minute maximum sustained winds exceeding 74 mph (119 km/h; 64 kn). Most of these storms occur between June 1 and November 30 each year, a time period referred to as the Atlantic hurricane season.

Atmosphere: The various layers of gases surrounding the Earth and held in place by gravity. The Earth's atmosphere is the origin of the weather phenomena studied in meteorology. Atmospheric composition, temperature, and pressure vary across a series of distinct sublayers including the troposphere and stratosphere.

Atmospheric: circulationThe global-scale movement of air masses within the Earth's atmosphere. All meteorological phenomena are a consequence of the atmospheric circulation, which manifests as a network of both latitudinal and longitudinal "cells" of convective activity; together with ocean circulation, these cells are the primary means by which thermal energy from the Sun is redistributed across the Earth's surface.

Atmospheric: soundingA measurement of the vertical distribution of physical properties through an atmospheric column, usually including pressure, temperature, wind speed and direction, moisture content, ozone concentration, and pollution, among others.

Backscatter: The diffuse reflection of waves, particles, or signals back to the same direction from which they originated. Backscattering is the principle underlying all weather radar systems, which can distinguish radar returns backscattered from target aerosols such as raindrops and snowflakes because the strength of the returns depends largely on the size and reflectivity of the targets.

Barometer: A scientific instrument used to measure atmospheric pressure.

Bernoulli's principle: A principle of fluid dynamics which states that an increase in the speed of a moving fluid occurs simultaneously with a decrease in the pressure exerted by the fluid or in the fluid's potential energy.

Blizzard: A severe snowstorm characterized by strong sustained winds of at least 35 mph (56 km/h) and typically lasting three hours or more. They can have an immense size, covering hundreds or thousands of square miles, and occur most often in temperate, polar, or mountainous regions during the winter.

Bow echo: A characteristic radar return from a mesoscale convective system that is shaped like an archer's bow and usually associated with squall lines or lines of convective thunderstorms. The distinct bow shape is a result of the focusing of a strong flow at the rear of the system. Especially strong bow echoes may develop into derechos.

Bulk Richardson Number (BRN): A dimensionless ratio related to the consumption of turbulence divided by the shear production of turbulence (the generation of kinetic energy caused by wind shear). It is an approximation of the Gradient Richardson Number.

Central dense overcast (CDO) The large, centralized, contiguous area of thunderstorms surrounding the rotational center of a strong tropical or subtropical cyclone. When a cyclone reaches sufficient intensity, a distinguishable eye may develop within the CDO. The strongest winds and heaviest rainfall are usually found beneath the coldest cloud tops in the CDO.

Cirrus (Ci): A genus of cloud characterized by thin, wispy, feather-like strands that appear white or light grey in color and form at very high altitudes, usually between 5 and 13.7 km (16,000 and 45,000 ft) above sea level. Cirrus clouds often develop from the outflow of cumulonimbus clouds in advance of fronts or thunderstorms, and therefore may indicate the imminent arrival of precipitation.

Clear ice: A type of solid precipitation which forms when relatively large drops of water are supercooled into a dense, transparent coating of ice without air or other impurities. It is similar to glaze and hard rime and, when formed on the ground, is often called black ice.

Climate: The statistics of weather in a given region over long periods of time, measured by assessing long-term patterns of variation in temperature, atmospheric pressure, humidity, wind,precipitation, and other meteorological variables. The climate of a particular location is generated by the interactions of the atmosphere, hydrosphere, cryosphere, lithosphere, and biosphere and strongly influenced by latitude, altitude, and local topography. Climates are often classified according to the averages or typical ranges of different variables, most commonly temperature and precipitation

ClimatologyAlso called climate science. A branch of the atmospheric sciences that studies climate, defined as weather conditions averaged over an extended to indefinite period of time. Climatology incorporates aspects of oceanography, geology, biogeochemistry, and the related field of meteorology to understand the long-term dynamics of climate-influencing phenomena and to produce climate models which can be used to estimate past climates and predict future climates.

Climate legislation: Legislation dealing with regulation of greenhouse gas emissions.

Climate resilience: The capacity for a socio-ecological system to: (1) absorb stresses and maintain function in the face of external stresses imposed upon it by climate change and (2) adapt, reorganize, and evolve into more desirable configurations that improve the sustainability of the system, leaving it better prepared for future climate change impacts, while still being able to benefit from it now.

Cloud An aerosol consisting of a visible mass of minute liquid droplets, frozen crystals or other particles suspended in the atmosphere. Water or various other chemicals may compose the droplets and crystals. On Earth, clouds are formed as a result of the saturation of an air mass when it is cooled to its dew point or when it gains sufficient moisture (usually in the form of water vapor) from an adjacent source to raise the dew point to the ambient temperature. There are many different types of clouds, which are classified and named according to their shape and altitude.

Cloud atlas: A pictorial key to the classification and nomenclature of clouds.

Cloud base: The lowest altitude of the visible portion of a cloud.

Cloud cover: The obscuration of all or part of the sky by clouds as observed from a particular location, or the specific fraction of the sky obscured by clouds as measured in oktas

Cloud typeAlso called a cloud genus: Any of a set of Latin names used to classify and identify clouds occurring in the troposphere, typically by characteristics such as their altitude, shape, and convective activity. A set of 10 or 12 traditional cloud types defined by the World Meteorological Organization and further subdivided into cloud species and cloud varieties is widely used in meteorology. Other classification systems have proposed many additional types.

cloudburstThe point of intersection of a trough and a ridge in the pressure pattern of a weather map. It generally takes the shape of a saddle in which the air pressure is slightly higher than that within the low-pressure regions but still lower than that within the anticyclonic zones.

Cold front: A type of front located at the leading edge of a cooler air mass as it replaces a warmer air mass. Cold fronts lie within a sharp surface trough of low pressure and the temperature difference between the air masses they separate can exceed 30 °C (86 °F). When enough moisture or instability is present, lines of rain or thunderstorms may accompany the boundary as it moves. In surface weather analysis, cold fronts are symbolized by a blue line with triangles pointing in the direction of travel.

Cumulus (Cu): A genus of cloud characterized by low-level "puffy" or "cotton-like" forms with flat bases (generally opaque white in color but sometimes with grey undersides), which occur individually or multiply in a variety of distinct subforms, usually at altitudes less than 2 km (6,600 ft) above sea level. Cumulus clouds normally produce little or no precipitation, but can develop into precipitation-bearing clouds such as cumulonimbus when influenced by atmospheric instability, moisture, and temperature gradients.

Cyclone: Any large-scale air mass characterized by inward spiraling winds which circulate around a strong center of low atmospheric pressure. Cyclones can form over land or water, can vary in size from mesocyclones such as tornadoes to synoptic-scale phenomena such as tropical cyclones and polar vortices, and may transition between tropical, subtropical, and extratropical phases. Contrast anticyclone.

Daytime: The period of the day between sunrise and sunset, during which any given point on the Earth experiences natural illumination from especially direct sunlight, known as daylight.

Dew: Liquid water droplets that commonly appear on thin, exposed surfaces in the morning or evening due to the condensation of atmospheric moisture on radiatively cooled surfaces. When temperatures are low enough, the water droplets freeze into ice particles known as frost.

Dobson unit (DU): A unit of measurement used to describe the quantity of a trace gas (primarily atmospheric ozone concentrations) present in a vertical column of the atmosphere. It is defined as the thickness (in units equivalent to 10 μm) of the layer of pure gas which would be formed if all of the gas molecules in the column could be collected on the surface at standard temperature and pressure.

Drought: Any prolonged period of below-average precipitation in a given region that results in shortages in the local water supply, whether of atmospheric, surface water, or ground water. Droughts can last for months or even years, and may be declared after as few as 15 days; annual or seasonal decreases in precipitation, such as dry seasons in the tropics, are sometimes called droughts, though a true drought is by definition abnormal or irregular. Drought conditions result from the confluence of a wide variety of climatic factors and may be exacerbated by hot temperatures; in turn, droughts may increase the likelihood of wildfires

Dry lightning: Lightning associated with a dry thunderstorm.

Dry thunderstorm also called a heat storm: A thunderstorm that produces thunder and lightning but in which most or all of its precipitation evaporates before reaching the ground. Dry thunderstorms occur necessarily in dry conditions, and their lightning strikes, sometimes referred to as dry lightning, are a major cause of wildfires

Dust storm Also called a duster or rendered duststorm: .A meteorological phenomenon characterized by very strong winds that blow dust-filled air over an extensive area. Dust storms arise when a gust front or other strong wind blows loose dirt, sand, and/or small rocks from a dry surface into the atmosphere, drastically reducing visibility. Though the term is sometimes restricted to storms occurring over normally arable land sduffering from drought, it is also used interchangeably with sandstorm and haboob.

Earth's atmosphere: The layer of gases surrounding the Earth and retained by the Earth's gravity

Extreme weather: Any weather that is unexpected, unusual, unpredictable, unseasonal, or especially severe (i.e. weather at the extremes of an historical distribution).

Firestorm: A very large wildfire or other conflagration which because of its intensity is able to create and sustain its own storm-force winds. Firestorms develop when a convective updraft of hot air rising from the burning area draws in strong wind gusts from all directions, which supply the fire with additional oxygen and thereby induce further combustion. They are often associated with flammagenitus clouds and fire whirls

Flash flood: Any flood which very rapidly inundates low-lying areas such as washes, rivers, dry lakes, and basins, especially one which recedes again in less than six hours. Flash flooding can be caused by heavy rain associated with severe weather, large amounts of meltwater from melting ice or snow, or the sudden collapse of a natural ice or debris dam.

Flash freezing: The process by which objects such as liquid hydrometeors are cooled below their freezing point very quickly, typically upon being subjected to extremely cold atmospheric temperatures or by making contact with a frozen surface.

Flood: An overflow of water which submerges land that is usually dry. Flooding may occur when water bodies such as rivers, lakes, or oceans escape their boundaries by overtopping or puncturing levees, or it may occur when precipitation accumulates on saturated ground more rapidly than it can either infiltrate or run off

Fog: A visible aerosol of minute water droplets or ice crystals that is suspended in the air at or near the Earth's surface. Fog is often considered a type of low-lying cloud and is heavily influenced by local topography, nearby bodies of water, and wind conditions

Freezing rain: Liquid droplets of rain that become supercooled while falling through a sub-freezing air mass and then freeze upon impact with any surface they encounter; the resulting glaze ice can accumulate to a thickness of several centimeters. Unlike mixed rain and snow, ice pellets, and hail, freezing rain exists entirely as a liquid until it hits a surface

Frost: A very thin layer of ice crystals on a solid surface, typically restricted to that which forms when water vapor in an atmosphere whose temperature is above freezing comes into contact with a surface whose temperature is below freezing. Frost may exhibit a great variety of forms.

Geostrophic wind: The theoretical wind that would result from an exact balance between the Coriolis force and the pressure gradient force (known as geostrophic balance). The true wind almost always differs from the geostrophic wind due to the influence of other forces such as friction from the ground

Global warming (GW): Usually: the warming trend over the past century or so; also: any period in which the temperature of the Earth's atmosphere increases; also the theory of such changes

GPS meteorology: A type of observational meteorology that interprets the effects of atmospheric properties such as total precipitable water vapor on the propagation of Global Positioning System (GPS) radio signals to derive information about the state of the local atmosphere.

Ground blizzard: A weather condition that occurs when loose snow or ice on the ground is lifted and blown into the air by strong winds. This can create low-visibility conditions even in the absence of precipitation.

Gust: A brief, sudden increase in the speed of the wind, usually lasting less than 20 seconds. Gusts are more transient than squalls and are followed by a lull or slowing of the wind speed. They are generally only reported by weather stations when the maximum wind speed exceeds the average wind speed by at least 10–15 knots (12–17 mph)

Hail: A type of solid precipitation that consists of balls or irregular lumps of ice usually 5–150 mm (0.20–5.91 in) in diameter, each of which is called a hailstone. Hail formation requires environments with strong, upward motion of air and low altitudes at which water freezes, which makes it possible within most thunderstorms. It is distinct from graupel and sleet or ice pellets

HailstormAlso called the Lower Atmosphere Severity Index: A weather index that measures the potential for dry, unstable air to contribute to the development of large or erratic wildland fires. The index derives from data on the stability and moisture content of the lower atmosphere and is calculated over three ranges of atmospheric pressure.

Heat index (HI): Also called the **apparent temperature, felt air temperature**, or **humiture:** A meteorological index that posits the apparent temperature perceived by the average human being who is exposed to a given combination of air temperature and relative humidity in a shaded area. For example, when the air temperature is 32 °C (90 °F) with 70% relative humidity, the heat index is 41 °C (106 °F).

Humidity: A measure of the amount of water vapor present in a parcel of air. By quantifying the saturation of the air with moisture, humidity indicates the likelihood of precipitation, dew, or fog occurring. The amount of water vapor needed to achieve full saturation increases as the air temperature increases. Three primary measurements of humidity are widely employed in meteorology: absolute, relative, and specific.

Hydrometeor: Any particulate of liquid or solid water within the atmosphere, encompassing all types of precipitation, formations due to condensation such as clouds and haze, and particles blown from the Earth's surface by wind such as blowing snow and sea spray.

Hydrometeorology: A branch of meteorology and hydrology that studies the transfer of water and energy between land surfaces and the lower atmosphere.

Hydrosphere: The combined mass of all solid, liquid, and gaseous forms of water found on, beneath, or above the surface of the Earth, including all oceans, lakes, streams, groundwater, atmospheric water vapor, snow, ice caps, and glaciers.

Hygrometer: A scientific instrument used to measure humidity.

Hygroscopy: A scientific instrument used to measure height or elevation, either by trigonometry or by the principle that atmospheric pressure influences the boiling point of liquids.

Ice: Water frozen into a solid state. Ice is abundant on Earth's surface and in the atmosphere and plays a major role in Earth's water cycle and climate. Its natural occurrence in weather phenomena takes many forms, including snowflakes, hail, frost, icicles, and ice spikes

Ice storm: A type of winter storm characterized by freezing rain which results in the accumulation of at least 6.4 millimetres (0.25 in) of ice on exposed surfaces

Jet stream: A narrow, fast-flowing, meandering air current primarily occurring in the upper part of the troposphere, at altitudes above 9 km (30,000 ft), and usually flowing from west to east. The Northern and Southern Hemispheres each have a polar jet and a subtropical jet; low-level jets and other types of jet streams can form under certain conditions.

Kyoto Protocol: A modification to the United Nations Framework Convention on Climate Change. See also post–Kyoto Protocol negotiations on greenhouse gas emissions.

La Niña: A weather phenomenon produced when a cold air mass moves across long expanses of warmer lake water, which causes the lowest layers of air to pick up warm water vapor from the lake, rise through the upper layers, freeze and then precipitate on the lake's leeward shores. In combination with orographic lift, the effect produces narrow but very intense bands of precipitation, especially snow, which can deposit at very high rates and result in very large amounts of snowfall over a region. The same effect can also occur over bodies of salt water, when it is termed ocean-effect or bay-effect snow.

Land breeze: The movement of a storm or other weather phenomenon over land after being over water.

Level of free convection (LFC): The altitude in the atmosphere at which the temperature of the environment decreases faster than the moist adiabatic lapse rate of a saturated air parcel at the same level. Air masses with one or many LFCs are potentially unstable and may develop convective clouds such as cumulonimbus.

Lifted index (LI): The difference in temperature between the ambient environment and an air parcel that is lifted adiabatically at a given pressure height within the troposphere, typically 500 hectopascals (0.49 atm). When the value of the lifted index is positive, the atmosphere at the given height is stable; when it is negative, the atmosphere is unstable.

Lightning: A naturally occurring electrostatic discharge during which two electrically charged regions of the atmosphere or ground temporarily equalize themselves, instantaneously releasing about a billion joules of energy across a wide range of the electromagnetic spectrum, from very hot plasma to brilliant flashes of light visible in the atmosphere. Lightning is often followed by its audible consequence, thunder, and is one of the distinguishing features of thunderstorms. Lightning phenomena are generally separated into three classes based on where they occur – either inside a single cloud, between two different clouds, or between a cloud and the ground – but many other observational variants have been recognized.

Lysimeter: An instrument used to measure the total amount of evapotranspiration that occurs within a certain area of the Earth's surface, usually by recording the amount of precipitation received by the area and the amount of moisture subsequently lost through the soil.

Mass flow: The movement of a fluid, such as an air mass, down a pressure or temperature gradient.

Mesosphere: The third major layer of the Earth's atmosphere, above the stratosphere and below the thermosphere. The lower boundary of the mesosphere varies between 50 and 65 km (31 and 40 mi) above the Earth's surface, depending on latitude and time of year.

Meteorology: A branch of the atmospheric sciences which seeks to understand and explain observable weather events, with a major focus on weather prediction. Meteorology uses variables familiar in chemistry and physics to describe and quantify meteorological phenomena, including temperature, pressure, water vapor, mass flow, and how these properties interact and change over time.

Mini-tornado: A fallacious term often used in news media to refer to damaging winds accompanying a thunderstorm, indifferently caused by tornadoes or microburst, on a small area

Monsoon:1. An abrupt seasonal wind reversal accompanied by corresponding changes in precipitation.2. Any seasonal change in atmospheric circulation and precipitation associated with the asymmetric heating of land and sea. In this context, the term is often used to refer specifically to the rainy phase of such a pattern, and in some places colloquially (and less correctly) to any locally very heavy but short-term rainfall

National Tornado Database: The official NOAA record of all known tornadoes within the United States from 1950 to present.

National Weather Service (NWS): The national meteorological agency of the United States, tasked with providing weather forecasts, warnings of severe weather, and other weather-related services to organizations and the public for the purposes of protection, safety, and general information

Nephoscope: A scientific instrument used to measure the altitude, direction, and velocity of atmospheric clouds.

Pascal (Pa): The SI derived unit of pressure, defined as one newton per square metre. In meteorology, measurements of atmospheric pressure are often given in kilopascals (kPa)

Photometeor: Any bright object or other optical phenomenon appearing in the Earth's atmosphere when sunlight or moonlight creates a reflection, refraction, diffraction, or interference under particular circumstances. Common examples of photometeors include halos, coronae, rainbows, crepuscular rays, and sun dogs

Polar vortex: Either of the two very large, persistent, rotating, upper-level low-pressure areas suspended in the Earth's atmosphere near the geographic poles. The polar vortices predictably strengthen during their local winter and weaken during their local summer as the temperature contrast between the poles and the Equator changes. When either vortex is weak, high-pressure zones of lower latitudes may push poleward, driving the vortex, jet stream, and masses of cold, dry polar air into the mid-latitudes, which can cause sudden, dramatic drops in temperature known as cold waves.

Precipitation: Any product of the condensation of atmospheric water vapor that falls by gravity, the main forms of which include rain, sleet, snow, hail, and graupel.

Precipitation occurs when a portion of the atmosphere becomes locally saturated with water vapor such that the water condenses into liquid or solid droplets and thus "precipitates" out of the atmosphere.

Prevailing winds: The predominant winds encountered at a particular point or region of the Earth's surface, identified by their source and direction. Though wind speed and direction can vary widely for a given location at a given time, the prevailing winds represent the primary trend in the characteristics of local winds averaged over a long period of time. They are influenced both by global patterns of atmospheric air movements and by local topography.

Quantitative precipitation estimation (QPE): A method of estimating the approximate amount of precipitation that has fallen at a location or across a region.

Radiosonde: A battery-powered scientific instrument released into the atmosphere, usually by a weather balloon, which measures various atmospheric parameters and transmits them by radio telemetry to a ground receiver. Radiosondes are essential sources of meteorological data.

Rain: A type of precipitation that occurs when liquid water in the form of droplets condenses from atmospheric water vapor, becoming heavy enough to fall under gravity. Rain is a major component of the water cycle and is responsible for depositing most of the fresh water on the Earth.

Rain gauge: An instrument used to collect and measure the amount of liquid precipitation that occurs within a certain area over a certain period of time.

Rain shadow: A relatively and consistently dry area on the leeward side of a significant geographic uplift such as a mountain range. Rain shadows exist because the mountains act as a barrier to the passage of precipitation-producing weather systems: moist air masses crossing areas of high elevation are forced upward by orographic lift, which causes the moisture to condense and precipitate on the windward side and leaves the air depleted of moisture by the time it reaches the leeward side.

Remote sensing: The acquisition of information about an object or phenomenon without making physical contact with the object and thus in contrast to on-site observation. In meteorology, satellite- or aircraft-based sensor technologies are widely used to detect and classify objects on the surface or within the atmosphere or oceans based on propagated electromagnetic signals.

Season: Any division of the year marked by changes in weather, ecology and the duration of daylight. Seasons result from the Earth's orbit around the Sun and its axial tilt relative to the ecliptic plane. In temperate and polar regions, four calendar-based seasons – spring, summer, autumn and winter – are generally marked by significant changes in the intensity of sunlight that reaches the Earth's surface; these changes become less dramatic as one approaches the Equator, and so many tropical regions have only two or three seasons, such as a wet season and a dry season. In certain parts of the world, the term is also used to capture the timing of important ecological events, such as hurricane seasons, flood seasons and wildfire seasons.

Snow: A type of solid precipitation in the form of ice crystals which precipitate from the atmosphere and subsequently undergo changes on the Earth's surface. Snow occurs when particles in the atmosphere attract supercooled water droplets, which nucleate and freeze into hexagonal crystals known as snowflakes; upon reaching the ground it may then accumulate into snowpack or snowdrifts and, over time, metamorphose by sintering, sublimation, and freeze-thaw mechanisms. Unless the local climate is cold enough to maintain persistent snow cover on the ground, snow typically melts seasonally.

Snowbelt: A region near the Great Lakes of North America where heavy snowfall in the form of lake-effect snow is particularly common.

Snowdrift: A deposit of snow sculpted by wind into a mound during a snowstorm.

Storm: Any disturbed state of an environment or atmosphere especially affecting the ground surface and strongly implying severe weather. Storms are characterized by significant disruptions to normal atmospheric conditions, which can result in strong wind, heavy precipitation, and/or thunder and lightning (as with a thunderstorm), among other phenomena. They are created when a center of low pressure develops within a system of high pressure surrounding it.

Storm shelter: A type of underground bunker designed to protect the occupants from violent severe weather, particularly tornadoes.

Stratosphere: The second major layer of the Earth's atmosphere, above the troposphere and below the mesosphere. The lower boundary of the stratosphere varies between 7 and 20 km (4.3 and 12.4 mi) above the Earth's surface, depending on latitude.

Temperature: A physical quantity expressing the thermal motion of a substance, such as a mass of air in the atmosphere, and proportional to the average kinetic energy of the random microscopic motions of the substance's constituent particles. Temperature is measured with a thermometer calibrated in one or more temperature scales: the Kelvin scale is the standard used in scientific contexts, but the Celsius and Fahrenheit scales are more commonly used in everyday contexts and for weather forecasting.

Terminal aerodrome forecast (TAF): A format for reporting current and forecast weather conditions, particularly as such information relates to aviation. Standard TAFs are issued by major civil airfields at least four times a day (every six hours) and generally apply to a 24- or 30-hour period and an area within approximately 8 kilometres (5.0 mi) from the center of an airport runway complex. TAFs complement and use similar encoding to METAR reports, but also take into account local geographic influences on weather.

Thunder: The sound produced as a result of the sudden thermal expansion of air within and surrounding the channel of a lightning discharge. This expansion creates an audible supersonic shock wave that, depending on the listener's distance from the source, can range from a sharp, loud crack (sometimes called a thunderclap or peal of thunder) to a deep, sustained rumble. Thunder is a defining feature of thunderstorms.

Tornado vortex signature (TVS): A rotation algorithm detected by weather radar that indicates the likely presence of a strong mesocyclone such as a tornado. Such signatures can be used to track the location and development of a tornadic rotation within a larger storm.

Tropopause: The boundary in the Earth's atmosphere between the troposphere and the stratosphere, on average situated approximately 17 km (11 mi) above equatorial regions and 9 km (5.6 mi) above the polar regions.

Troposphere: The lowest layer of the Earth's atmosphere, within which nearly all weather phenomena occur. The troposphere contains approximately 75% of the atmosphere's total mass and 99% of its water vapor and aerosols. The average height of the troposphere above the Earth's surface varies between 6 and 18 km (3.7 and 11.2 mi) depending on latitude.

Typhoon: The local name for a tropical cyclone that occurs in the northwestern Pacific Ocean, between 180° and 100°E in the Northern Hemisphere.

Urban heat island: Any metropolitan area which is significantly warmer than its surroundings.

Urban heat island (UHI): An urban or metropolitan area within which air temperatures are significantly warmer than in surrounding rural or uninhabited areas as a result of human activities, especially the artificial modification of land surfaces and the generation of waste heat by energy usage. Urban heat islands can greatly influence precipitation, air quality, and the likelihood of certain weather phenomena in the vicinity of large cities, though not all cities have a distinct urban heat island.

Virtual temperature (Tv): The temperature of a moist air parcel at which a theoretical dry air parcel would have a total pressure and density equal to those of the moist parcel.

Water vapor: Water in its gaseous state. Water vapor is ubiquitous in the atmosphere, being continuously generated by evaporation and removed by condensation, and plays a major role in numerous meteorological processes.

Weather: The state of the atmosphere at a given time and location. Weather is driven by a diverse set of naturally occurring phenomena, especially air pressure, temperature, and moisture differences between one place and another, most of which occur in the troposphere.

Weather balloon: Also called a sounding balloon.: A high-altitude balloon used to carry scientific instruments into the atmosphere, which then measure, record, and transmit information about meteorological variables such as atmospheric pressure, temperature, humidity, and wind speed by means of a radiosonde or other measurement device, often one which is expendable.

Weather forecasting: The application of science and technology to predict the conditions of the atmosphere at a given time and location. Weather forecasts are made by collecting quantitative data about the current state of the atmosphere at a given place and then using meteorology to project how the atmosphere will change. Forecasting is important to a wide variety of human activities, including business, agriculture, transportation, recreation and general health and safety, because it can be used to protect life and property.

Weather map: A map which displays various meteorological features across a particular area for a particular point or range of time. Weather maps often use symbols such as station models to conveniently present complicated meteorological data. They are used for both research and weather forecasting purposes.

Weather station: Any facility, either on land or at sea, with instruments and equipment for measuring atmospheric conditions in order to provide information for weather forecasts and to study the weather and/or climate.

Wind: The bulk movement of air within the Earth's atmosphere. Wind occurs on a wide range of scales, from very strong thunderstorm flows lasting tens of minutes to milder local breezes lasting a few hours to global atmospheric circulations caused by the differential heating of the Equator and the poles and the Earth's rotation. Winds are often referred to by their strength and direction; the many types of wind are classified according to their spatial scale, their speed, the types of forces that cause them, the regions in which they occur, and their effects.

Wind gust: A brief increase in the speed of the wind, usually lasting less than 20 seconds. Gusts are more transient than squalls. They are usually only reported by weather stations when the maximum or peak wind speed exceeds the average wind speed by 10–15 knots (12–17 mph).

Wind speed: The measured speed of the air comprising a wind. Changes in wind speed are often caused by air parcels being exposed to pressure and temperature gradients in the atmosphere. Wind speed is measured with an anemometer, but may also be less precisely classified using the Beaufort scale.

Windstorm: Any storm that produces or is characterized by very strong winds.

REFERENCES

*Bluestein, Howard B. Synoptic-Dynamic Meteorology in Midlatitudes. Vol. 1: Principles of Kinematics and Dynamics, Oxford University Press, New York, 1992.

*Cotton, W. R., and R. A. Anthes. Storm and Cloud Dynamics, Academic Press, New York, 1989.

*Hobbs, Peter V. Basic Physical Chemistry for Atmospheric Sciences, Cambridge University Press, New York, 1995.

*Holton, James R. An Introduction to Dynamic Meteorology (2nd ed.), Academic Press, New York, 1979.

*Ray, Peter S., Ed. Mesoscale Meteorology and Forecasting, American Meteorological Society, Boston, MA, 1986.

*Rogers, R. R. A Short Course in Cloud Physics (3rd ed.), Pergamon Press, Oxford, England, 1989.

*Wallace, J. M., and P. V. Hobbs. Atmospheric Science: An Introductory Survey, Academic Press, New York, 1977.

*Watson, Robert T., et al., Eds. Climate Change 1995. Impacts, Adaptations and Mitigation of Climate Change: Scientific-Technical Analyses, Cambridge University Press, Cambridge, England, 1996.

A Half Century Of Progress In Meteorology,A Tribute To Richard Reed,Johnson, Richard H.January 2003.

A Tribute To Fred Sanders,Bosart, Lance F.January 2008.

Anthes, R. A. Tropical Cyclones: Their Evolution, Structure, and Effect, American Meteorological Society, Boston, MA, 1982.

Arya, Pal S. Air Pollution Meteorology, Oxford University Press, New York, 1998.

Asian Development Bank, 2009. Addressing Climate Change in the Asia and Pacific Region.

Austin Bourke, P.M, 1957: The Use of Synoptic Weather Maps in Potato Blight Epidemiology. Technical Note No. 23. Dublin, Irish Meteorological Service.

Baier, W. and G.W. Robertson, 1965: Estimation of latent evaporation from simple weather observations. Can. J. Plant Sci., 45:276–284.

Battan, Louis J. Harvesting the Clouds, Anchor Books, Doubleday and Co., Garden City, NY, 1961. Baumgardner, Darre l(2018),Ice Formation And Evolution In Clouds And Precipitation:Measurement And Modeling Challenges,Baumgardner, Darrel

Beaumount, A., 1947: The dependence on weather of the dates of potato blight epidemics. Trans. Br. Mycol. Soc., 31:45–53.

Bigg, Grant R. The Oceans and Climate, Cambridge University Press, New York, 1996.

Bohren, Craig F. Clouds in a Glass of Beer: Simple Experiments in Atmospheric Physics, Wiley, New York, 1987.

Boken, V.K, A.P. Cracknell and R.L. Heathcote (eds), 2005: Monitoring and Predicting Agricultural Drought: A Global Study. New York, Oxford University Press.

Bouma, E. and J.G. Hansen, 1999: Overview of standard descriptions of Phytophthora Decision Support Systems. In: Proceedings of the Workshop on the European Network for Development of an Integrated Control Strategy of Potato Blight. Uppsala, Sweden, 4–13 September 1998 (H. Schepers and E. Bouma, eds). Lelystad, Pub. Applied Research for Arable Farming and Field Production of Vegetables.

Briefing Note (1), 2008, International Strategy for Disaster Reduction, Geneva, September, 2008 International Food Policy Research Institute, 2009. Climate Change: Impact on Agriculture and Costs of Adaptation, 2009 Jamil Ahmad, Dastgir Alam and Ms. Shaukat Haseen. 2011.

Cotton, William R. Storms, ASTeR Press, Fort Collins, CO, 1990.

Cotton, William R., and Roger A. Pielke. Human Impacts on Weather and Climate, Cambridge University Press, New York, 1995.

Daly, C., R.P. Neilson and D.L. Phillips, 1994: A statistical-topographic model for mapping climatological precipitation over mountainous terrain. J. Appl. Meteor., 33: 140–158.

De Blij, H. J. Nature on the Rampage, Smithsonian Books, Washington, DC, 1994.

Document on National Action Plan on Climate Change, India IPCC (Intergovernmental Panel on Climatic Change) 2006. The Economics of Climate Change: Stern Review. The Summary of conclusions. Survey of the Environment 2007, The Hindu, pp141-145.

Dyras, I., H. Dobesch, E. Grueter, A. Perdigao, O.E. Tveito, J.E. Thornes, F. van der Wel and L. Bottai, 2005. The use of Geographic Information Systems in climatology and meteorology: COST 719. Meteorol. Appl., 12:1–5.

Eagleman, Joe R. Air Pollution Meteorology, Trimedia Publishing Co., Lenexa, KS, 1991.

Elsner, James B., and A. Biral Kara. Hurricanes of the North Atlantic, Oxford University Press, New York, 1999.

Elsom, Derek M. Atmospheric Pollution: A Global Problem (2nd ed.), Blackwell Publishers, Oxford, England, 1992.

Encyclopedia of Climate and Weather, Vol. 1 and Vol. 2, Stephen H. Schneider, Ed., Oxford University Press, New York, 1996.

Energy and Climate Change. Report of the DOE Multi-Laboratory Climate Change Committee, Lewis Publishers, Chelsea, MI, 1991.

England, Gary A. Weathering the Storm, University of Oklahoma Press, Norman, OK, 1996.

Essentials Of Meteorology: An Invitation To The Atmosphere [With Cdrom] (Paperback) By C. Donald Ahrens (Shelved 4 Times As Meteorology)Avg Rating 3.64 — 117 Ratings — Published 1962

Fedoseev, A.P., 1979: Field Management and Weather. Leningrad, Hydrometeoizdat.
———, 1985: Agrometeorological Conditions for the Effectiveness of Fertilization. Leningrad, Hydrometeoizdat.

Folkedal, A. and C. Brevig, 2004: VIPS - A web-based decision support system for crop protection in Norway. In: Irrigation and Pest and Disease. , 1998: Crop Evapotranspiration. Guidelines for Computing Crop Water Requirements (R.G. Allen, L.S. Pereira, D. Raes and M. Smith). Irriggation and Drainage Paper No. 56. Rome, FAO.

Forsund, E., 1983: Late blight forecasting in Norway 1957–1980. EPPO Bulletin, 13:255–258.

Friesland, H., 2005: Description and testing of the model ASCHORF for apple scab infections. In: Irrigation and Pest and Disease Models: Evaluation in Different Environments and Web-based Applications (G. Maracchi, L. Kajfez-Bogataj, S. Orlandini, F. Rossi and M. Barazutti, eds). COST Action 718: Meteorological Applications for Agriculture. Brussels, European Commission.

Friesland, H., S. Orlandini and A. Sušnik, 2005: Description and testing of the model PERO for grapevine downy mildew predictions. In: Irrigation and Pest and Disease Models: Evaluation in Different Environments and Web-based Applications (G. Maracchi, L. Kajfez-Bogataj, S. Orlandini, F. Rossi and M. Barazutti, eds). COST

Action 718: Meteorological Applications for Agriculture. Brussels, European Commission.

Gloster, J., H.J. Champion, J.H. Sorensen, T. Mikkelsen, D.B. Ryall, P. Astrup, S. Alexandersen and A.I. Donaldson, 2003: Airborne transmission of foot-and-mouth disease virus from Burnside Farm, Heddon-onthe-Wall, Northumberland, during the 2001 epidemic in the United Kingdom. Veterinary Record, 152:525–533.

Gloster, J., R.M. Blackall, R.F. Sellers and A.I. Donaldson, 1981: Forecasting the airborne spread of foot-and-mouth disease. Veterinary Record, 110:47–52.

Gordeev, A.V., A.D. Kleschenko, B.A. Chernyakov and O.D. Sirotenko, 2006: Bioklimaticheskii potentsial Rossii: teoriia i praktika. [Bioclimatic Potential of Russia: Theory and Practice]. Moscow, KMK Scientific Press.

Greenler, Robert. Rainbows, Halos and Glories, Cambridge University Press, New York, 1980.

Hahn, D. and J.D. McQuigg, 1970: Evaluation of climatological records for rational planning of livestock shelters. Agric. Meteorol., 7:131–141.

Hansen, J.G., 1999: NegFry – A Decision Support System for Scheduling the Chemical Control of Potato Late Blight. User Manual. Tjele, Danish Institute of Agricultural Sciences.

Hansen, R.C., H.M. Keener and R.J. Gustafson, 1990: Natural Air Grain Drying in Ohio. Extension Factsheet AEX-202-90. Columbus, Ohio State University.

Her Majesty's Stationery Office (HMSO), 1967: Potential Transpiration. Technical Bulletin No.16. London, Ministry of Agriculture, Fisheries and Food.

Hodges, T. (ed.), 1991: Predicting Crop Phenology. Boca Raton, CRC Press.

Hogg, W.H., 1967: Atlas of Long-term Irrigation Needs for England and Wales. London, Ministry of Agriculture, Fisheries and Food.

Hunter, R.D. and R.K. Meentemeyer, 2005: Climatologically aided mapping of daily precipitation and temperature. J. Appl. Meteor., 44:1501-1510.

Hyre, R.A., 1954: Progress in forecasting late blight of potato and tomato. Plant Disease Reporter, 38:245–253.

Impact of Climate Change on Agriculture and Food Security in India Int. Jr. of Agril., Env. and Biotech. Vol. 4, No. 2 : June 2011 : 129-137.

International Cloud Atlas. World Meteorological Organization, Geneva, Switzerland, 1987.

IPCC (Intergovernmental Panel on Climatic Change) 2007. Climate Change: The Physical Science Basis. Extracts from the IV Assessment Report. Survey of the Environment 2007, The Hindu, pp147-155

James, Bruce P., et al., Eds. Climate Change 1995. Economic and Social Dimensions of Climate Change, Cambridge University Press, Cambridge, England, 1996.

Johnson, D.A., J.R. Alldredge and D.L. Vakoch, 1996: Potato late blight forecasting models for the semiarid environment of south-central Washington. Phytopathology, 86:480–484.

Keane, T., 1982: Weather and Potato Blight. Agrometeorological Memorandum No. 8. Dublin, Irish Meteorological Service.

Krishnan, A. and Singh, M. (1968). January. Soil climatic zones in relation to cropping patterns. In Proceedings of the Symposium on Cropping Patterns, Indian Council of Agricultural Research, New Delhi (pp. 172-185).

Kroes, J.G., 2005: An introduction to the SWAP model. In: Irrigation and Pest and Disease Models: Evaluation in Different Environments and Webbased Applications (G. Maracchi, L. Kajfez-Bogataj, S. Orlandini, F. Rossi and M. Barazutti, eds). COST Action 718: Meteorological Applications for Agriculture. Brussels, European Commission.

Leedman, A., 2007. The Australian National Agricultural Monitoring System: A national climate risk management application. In: Management of Natural and Environmental Resources for Sustainable Agriculture (R. Stefanski and P. Pasteris, eds). Proceedings of the USDA/ WMO Workshop on Management of Natural and Environmental Resources for Sustainable Agricultural Development, Portland, Oregon 13–16 February 2006. Technical Bulletin No. WAOB-2008, NRCS-2008 and AGM-10 (WMO/ TD-No.1428). Washington, DC, USDA; Geneva, WMO.

Madelin, M. and G. Beltrando, 2005: Spatial interpolation-based mapping of the spring frost hazard in the Champagne vineyards. Meteorol. Appl., 12:51–56.

Magnus, H.A., K. Munthe, E. Sundheim and A. Ligaarden, 1991: PC-technology in plant protection warning systems in Norway. Danish J. Plant Soil Sci., 85(S–2161): 1–6.

Marshall, J.K., 1967: The effect of shelter on the productivity of grasslands and field crops. Field Crop Abs., 20:1–14.

McGrath, R. and K. Finkle, 2001: HIRLAM Foot-and Mouth Disease Dispersion Model. Hirlam Progress Report 21. Dublin, Met Éireann.

Mercer, P.C., A. Bell, L.R. Cooke, L. Dowley, B. Dunne, T. Keane, T. Kennedy and R. Leonard, 2001: Crop and animal disease forecasting and control: regional perspectives. In: AgroMeteorological Modelling: Principles, Data and Applications (N.M. Holden, ed.). Dublin, Agmet.

Mesoscale Meteorology And Forecasting Ray, Peter S.January 1986

Meteorology Today: An Introduction To Weather, Climate, And The Environment [With Meteorologynow & Infotrac] (Hardcover)By C. Donald Ahrens (Shelved 4 Times As Meteorology)Avg Rating 3.97 — 153 Ratings — Published 1982

Meteorology Today: An Introduction To Weather, Climate, And The Environment, 9th Edition Harleigh Lu

Mikkelsen, T., S. Alexandersen, P. Astrup, H.J. Champion, A.I. Donaldson, F.N. Dunkerley, J. Gloster, J.H. Sørensen and S. Thykier-Nielsen, 2003: Investigation of airborne foot-and-mouth disease virus transmission during low-wind conditions in the early phase of the UK 2001 epidemic. Atmos. Chem. Phys., 3:2101–2110.

Mikkelsen, T., S.E. Larsen and S. Thykier-Nielsen, 1984: Description of the Risø puff diffusion model. Nucl. Technol., 67:56–65.

Misra. H.P. Changes in Crop Pest Scenario in Climate Change Context, Professor, Department of Entomology, College of Agriculture. O.U.A.T., Bhubaneswar

Models: Evaluation in Different Environments and Web-based Applications (G. Maracchi, L. KajfezBogotaj, S. Orlandini, F. Rossi and M. Barazutti, eds). COST Action 718: Meteorological Applications for Agriculture. Brussels, European Commission. Food and Agriculture Organization of the United Nations (FAO), 1977: Guidelines for Predicting Crop Water Requirements (J. Doorenbos and W.O. Pruitt). FAO Irrigation and Drainage Paper No. 24, Revised. Rome, FAO.

Motha, R. and R. Stefanski, 2006: United States Department of Agriculture's weather and climate information system for operational applications in agriculture. Meteorol. Appl., 13(1):31–47.

Motha, R. and T. Heddinghaus, 1986. The Joint Agricultural Weather Facility's Operational Assessment Program. Bull. Amer. Meteor. Soc., 67:1114–1122.

Motroni A., P. Duce, D. Spano and S. Canu, 2002: Estimation of climatic risk for agriculture in a Mediterranean region. In: Proceedings of the 15th Conference on Biometeorology and Aerobiology Joint with the 16th International Congress on Biometeorology of the International Society of Biometeorology, Kansas City, Missouri, 28 October–1 November 2002.

Mukhala, E., 2005: Food and Agriculture Organization and agricultural drought. In: Monitoring and Predicting Agricultural Drought: A Global Study (V.K. Boken, A.P. Cracknell and R.L. Heathcote, eds). New York, Oxford University Press.

Murphy, T.M., D.J. O'Brien and J.S. Gray, 2004: Weather, pests and epidemiology of animal diseases. In: Climate, Weather and Irish Agriculture. Second edition. (T. Keane and J.F. Collins, eds). Dublin, Agmet.

National Academy Press. Acid Deposition—Atmospheric Processes in Eastern North America, Washington, DC, 1983.

Negre, T., 2006: EU/JRC Agrometeorological monitoring and national services: opportunities and challenges. In: Strengthening Operational Agrometeorological Services at the National Level. Proceedings of the Inter-Regional Workshop, Manila, Philippines, 22–26 March 2004. (R. Motha, M.V.K. Sivakumar and M. Bernardi, eds). Technical Bulletin No. WAOB–2006–1 and AGM-9 (WMO/TD-No. 1277). Washington, DC, USDA; Geneva, WMO; Rome, FAO.

Ollrenshaw, B., 1966: The approach to forecasting the incidence of fascioliasis over England and Wales, 1958–1962. Agric. Meteorol., 3:35-53.

Orlandini, S., 1996: Agrometeorological models for crop protection. In: Proceedings of the International Symposium on Applied Agrometeorology and Agroclimatology, Volos, Greece, 24–26 April 1996. COST 77, 79, 711. Brussels, European Commission.

Orlandini, S., A.D. Marta, H. Friedland and A. Sušnik, 2005: Description and testing of PLASMO model for the simulation of grapevine downy mildew. In: Irrigation and Pest and Disease Models: Evaluation in Different Environments and Webbased Applications (G. Maracchi, L. Kajfez-Bogataj, S. Orlandini, F. Rossi and M. Barazutti, eds). COST Action 718: Meteorological Applications for Agriculture. Brussels, European Commission.

Pedgley, D., 1982: Windborne Pests and Diseases: Meteorology of Airborne Organisms. New York, John Wiley and Sons.

Petr, J., 1991: Weather and Yield. Developments in Crop Science 20. Amsterdam, Elsevier.

Prasad, R and Rana, R. 2006. A study on maximum temperature during March 2004 and its impact on rabi crops in Himachal Pradesh. J. of Agrometeorology, 8(1): 91-99

Prasada Rao, G.S.L.H.V. and Alexander, D. 2007. Impact of climate change on the agricultural sector in tropical countries. Proceedings of the WTO Workshop, held at College of Fisheries, Panangad, Kochi on 14th December, 2007, Kerala Agricultural University, 80p

Prasada Rao, G.S.L.H.V. and Alexander, D. 2008. Impact of climate change on the agricultural sector in tropical countries-. Recommendations. Proceedings of the WTO Workshop, held at College of Fisheries, Panangad, Kochi on 14th December, 2008, Kerala Agricultural University, 26p

Prasada Rao. G. S. L. H. V. impact of weather extremes on indian food grains Production

Pruess, K.P., 1983: Day-degree methods for pest management. Environ. Entomol.,12: 613–619.

Puterbaugh, T., R. Stefanski and M. Brusberg, 1997: The Joint Agricultural Weather Facility's operational procedures for processing and analyzing global crop and weather data. In: Proceedings of the 13th Conference on Interactive Information and Processing Systems for Meteorology, Oceanography, and Hydrology, Long Beach, California, 2–7 February 1997. Boston, American Meteorological Society.

Ramakrishna, Y.S., Rao, G.G.S.N., Rao, S.G. and Vijayakumar, P. 2006. Impact of climate change in Agriculture. In: Environment and Agriculture (eds. Chadha, K.L. and Swaminathan, M.S.). Malhotra Publishing House, New Delhi, pp. 1-30

Righter, Robert W. Wind Energy in America, University of Oklahoma Press, Norman, OK, 1996.

Rijks, D. and M.W. Baradas, 2000: The clients for agrometeorological information. Agric. For. Meteorol., 103:27–42.

Rosenberg, N.J., B.L. Blad and S.B. Verma, 1983: Microclimate: The Biological Environment. Second edition. New York, John Wiley and Sons.

Rossi, F., S. Orlandini, M. Magli and G. Maracchi, 2005: PLASMO Project: an Internet application. In: Irrigation and Pest and Disease Models: Evaluation in Different Environments and Web-based Applications (G. Maracchi, L. KajfezBogataj, S. Orlandini, F. Rossi and M. Barazutti, eds). COST Action 718: Meteorological Applications for Agriculture. Brussels, European Commission.

Rowland, J., J. Verdin, A. Adoum and G. Senay, 2005: Drought monitoring techniques for famine early warning systems in Africa. In: Monitoring and Predicting Agricultural Drought: A Global Study (V.K. Boken, A.P. Cracknell and R.L. Heathcote, eds). New York, Oxford University Press.

Rusakova, T.I., V.M. Lebedeva, I.G. Gringof and N.M. Shklyaeva, 2006: Modern technology of stage-by-stage forecasting of crop yield and gross collection. Meteorol. Hydrol., 7:101–108.

Ryall, D.B. and R.H. Maryon, 1998: Validation of the UK Met. Office's NAME model against the ETEX dataset. Atmos. Environ., 32:4265–4276.

Samra, J.S., Singh, G and Ramakrishna, Y.S. 2004. Cold wave during 2002-03 over North India and its effect on crops. The Hindu dated 10th January, 2004. p. 6.

Scorer, Richard S., and Arjen Verkaik. Spacious Skies, David and Charles Publishers, London, 1989.

Sentelhas, P.C., T.J. Gillespie, M.L. Gleason, J.E.B.A. Monteiro and S.T. Helland, 2004: Operational exposure of leaf wetness sensors. Agric. For. Meteorol., 126:59–72.

Shukla, P.R., Sharma, S.K and Ramana, V.P. 2002. Climate Change and India - Issues, Concerns and Opportunities. Tata – McGraw-Hill Publishing Company Ltd, New Delhi, 314p

Simpson, Robert H. and Herbert Riehl. The Hurricane and Its Impact, Louisiana State University Press, Baton Rouge, LA, 1981.

Singh. H.S. Potential Impact of Climate Change on Mangroves in India (0894-B2)

Smith, C.V., 1964: Animal housing and meteorology. I: Ventilation and associated pattern of air flow; II: The rating of ventilation systems for animal houses; III: A quantitative relationship between environment, comfort and animal productivity. Agric. Meteorol., 1: [page numbers unknown].

Smith, L.P. and M.E. Hugh-Jones, 1969: The effects of wind on the spread of foot-and-mouth disease. Nature, 223:712–715.

Smith, L.P., 1956: Potato blight forecasting by 90 per cent humidity criteria. Plant Pathology, 5:83–87.

Smith, L.P., 1956: Potato blight forecasting by 90 per cent humidity criteria. Plant Pathology, 5:83–87.

Smith, M., 2000: The application of climate data for planning and management of sustainable rainfed and irrigated crop production. Agric. For. Meteorol., 103:99–108.

Sørensen, J.H., 1998: Sensitivity of the DERMA long-range dispersion model to meteorological input and diffusion parameters. Atmos. Environ., 32:4195–4206.

Stull, Roland B. Meteorology Today for Scientists and Engineers, West Publishing Co., Minneapolis, MN, 1995.

Synoptic-Dynamic Meteorology And Weather Analysis And Forecasting

The Weather Factor,Ludlum, David McwilliamsJanuary 1984

Ullrich, J. and H. Schrodter, 1966: Das Problem der Vorhersage des Aufretens der Kartoffelkrautfaule (Phytophthora infestans) und die Möglichkeit seiner Losung durch eine Negativprognose. Nachrichtenblatt Dt. Pflanzenschutzdienst (Braunschweig), 18:33–40.

van Eimern, J., 1968: Problems of shelter planning. In: Agroclimatological Methods: Proceedings of the Reading Symposium. Paris, UNESCO.

van Everdingen, E., 1926: Het verband tusschen de weergesteldheid en de aardappelziekte (P. infestans). Tijdschrift Plantenziekten, 32:129–140.

van Rij, N.C. and E.D. du Preez, 2004: Use of empirical rules in the development and implementation of a potato late blight (Phytophthora infestans) incidence map for KwaZulu-Natal. Presented at 42nd Annual Congress of the Southern African Society for Plant Pathology, Underberg, KwaZulu-Natal, 18–21 January 2004. http://www.saspp.org/ abstracts2004/SASPP2004.pdf.

Venäläinen, A., T. Salo and C. Fortelius, 2005: The use of numerical weather forecast model predictions as a source of data for irrigation modeling. Meteorol. Appl., 12:307–318.

Volvach, V.V., 1987: Simulation of the Effects of Agrometeorological Conditions on the Development of Colorado Beetle. Leningrad, Hydrometeoizdat.

Weather (Paperback)By Storm Dunlop(Shelved 3 Times As Meteorology)Avg Rating 3.79 — 19 Ratings — Published 1999

Weathering The Storm,Sverre Petterssen, The D-Day Forecast, And The Rise Of Modern Meteorology,Fleming, James Rodger,January 2001

Williams, Jack. The USA Today Weather Book, Vintage Books, Random House, New York, 1992.

Winstel, K., 1993: Kraut- und Knollenfaule der Kartoffel eine neue Prognosemöglichkeit-sowie Bekampfungsstrategien. Med. Fac. Landbouww. Univ. Gent, 58(3b):1477–1483.

World Meteorological Organization, (2001). Lecture Notes for Training Agricultural Meteorological Personnel (J. Wieringa and J. Lomas). Second edition (WMO-No. 551), Geneva.

World Meteorological Organization, 1955: The Forecasting from Weather Data of Potato Blight and Other Plant Diseases and Pests (P.M. Austin Bourke). Technical Note No. 10 (WMO-No. 42), Geneva.

Wright, P.B., 1969: Effects of wind and precipitation on the spread of foot-and-mouth disease. Weather, 24:204–213.

Zhukov, V.A., A.N. Polevoy, A.N. Vitchenko and S.A. Daniyelov, 1989: Mathematical Methods for the Assessment of Agroclimatic Resources. Leningrad, Hydrometeoizdat.

Zoidze, E.K. and L.I. Ovcharenko, 2000: Comparative Assessment of the Agricultural Potential of Russian Climate and the Degree of Utilization of its Agroclimatic Resources by Crops. St Petersburg, Hydrometeoizdat.

INDEX

www.ingramcontent.com/pod-product-compliance
Lightning Source LLC
LaVergne TN
LVHW062228200926
848484LV00006B/1499